COMPLEX NUMBERS

NCEA Level 3 External

Charlotte Walker and Victoria Walker

Walker Maths 3.5 Complex Numbers
1st Edition
Charlotte Walker
Victoria Walker

Cover design: Cheryl Rowe, Macarn Design
Text design: Cheryl Rowe, Macarn Design
Production controller: Siew Han Ong

For product information and technology assistance,
in Australia call **1300 790 853**;
in New Zealand call **0800 449 725**

For permission to use material from this text or product, please email **aust.permissions@cengage.com**

National Library of New Zealand Cataloguing-in-Publication Data
A catalogue record for this book is available from the National Library of New Zealand

978 0 17 044705 8

Cengage Learning Australia
Level 7, 80 Dorcas Street
South Melbourne, Victoria Australia 3205

For learning solutions, visit **cengage.co.nz**

Printed in China by 1010 Printing International Limited.
4 5 6 7 24

CONTENTS

ISBN: 9780170447058

 ISBN: 9780170447058

Glossary

Make your own glossary of key terms:

Term	Definition	Picture/Example
Roots		
Discriminant		
Surd		
Conjugate surd		
Imaginary numbers		
Real numbers		
Complex numbers		
Re(z)		
Im(z)		
Arg(z)		

ISBN: 9780170447058

Term	Definition	Picture/Example
$\bar{z}$		
$\lvert z \rvert$		
Modulus		
Argument		
Argand plane		
Cartesian plane		
Rectangular form		
Polar form		
Cis		
Locus		

ISBN: 9780170447058

Revision and useful background

1 Polynomial equations

Expanding three brackets

Steps:

$(x + 3)(2x - 5)^2 = (x + 3)(2x - 5)(2x - 5)$ — If necessary, write as three sets of brackets.

$= (x + 3)(4x^2 - 10x - 10x + 25)$ — Multiply any two of the sets of brackets using **FOIL**.

$= (x + 3)(4x^2 - 20x + 25)$ — Collect the like terms.

$= 4x^3 - 20x^2 + 25x + 12x^2 - 60x + 75$ — Multiply the result by each term in the first set of brackets.

$= 4x^3 - 8x^2 - 35x + 75$ — Collect the like terms.

Example:

$$(3x - 2)(x + 2)(2x - 5) = (3x - 2)(2x^2 - 5x + 4x - 10)$$
$$= (3x - 2)(2x^2 - x - 10)$$
$$= 6x^3 - 3x^2 - 30x - 4x^2 + 2x + 20$$
$$= 6x^3 - 7x^2 - 28x + 20$$

Expand and simplify these.

1 $(x - 1)(x + 4)(x - 3)$

2 $(x + 5)(x - 3)(2x - 1)$

3 $(y + 1)(3y - 1)(2y + 3)$

4 $(p + 4)^2(2p - 3)$

5 $(4y - 5)^3$

6 $(x^2 + 3y)^3$

ISBN: 9780170447058

Solving using the quadratic formula

- Note that you **must** be able to use this formula. You will not always be able to rely on your calculator.

$ax^2 + bx + c = 0$

a is the coefficient of x^2

b is the coefficient of x

c is the constant

$$x = \frac{-b \pm \sqrt{b^2 - 4ac}}{2a}$$

Examples:

1 $9x^2 - 30x + 25 = 0$

$a = 9, b = -30, c = 25$

$$x = \frac{-(-30) \pm \sqrt{(-30)^2 - 4 \times 9 \times 25}}{2 \times 9}$$

$$= \frac{30 \pm \sqrt{0}}{18}$$

$$= \frac{5}{3}$$

One solution only.

2 $x(x + 5) = 1$

Expand and rearrange first.

$x^2 + 5x = 1$

$x^2 + 5x - 1 = 0$

$a = 1, b = 5, c = -1$

$$x = \frac{-5 \pm \sqrt{5^2 - 4 \times 1 \times (-1)}}{2 \times 1}$$

$$= \frac{-5 \pm \sqrt{29}}{2}$$

$$= -\frac{5}{2} \pm \frac{1}{2}\sqrt{29}$$

You will often be asked to leave your answer in this form.

Solve these using the quadratic formula. Where appropriate, leave your answer in the form $a \pm b\sqrt{c}$.

1 $x^2 + 4x - 6 = 0$

2 $4x^2 - 12x + 9 = 0$

3 $3x^2 + 7x = 2$

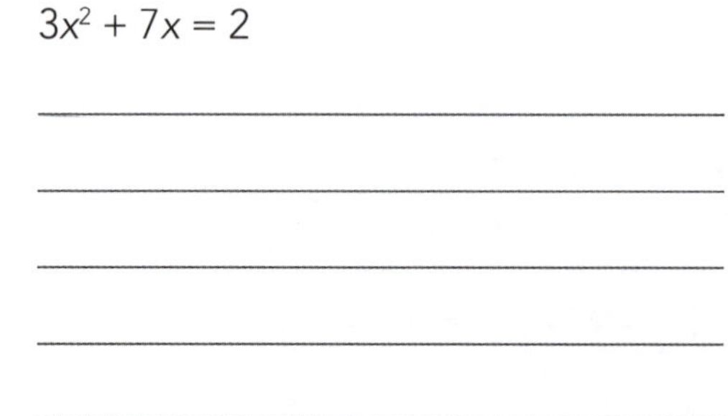

4 $11x^2 = 5 + 8x$

 ISBN: 9780170447058

Roots of equations and use of the discriminant

Roots of equations

- Quadratic equations have a **maximum** of **two solutions**, which represent points where the parabola crosses the *x*-axis. These solutions are also known as **roots**.
- So far you have dealt only with 'real' roots. However, in this standard you will learn about **imaginary** or **unreal** roots.
- The value of $b^2 - 4ac$ in the quadratic formula tells you how many roots there are. This value is known as the discriminant (Δ).
- There are three possibilities for the number of real roots:

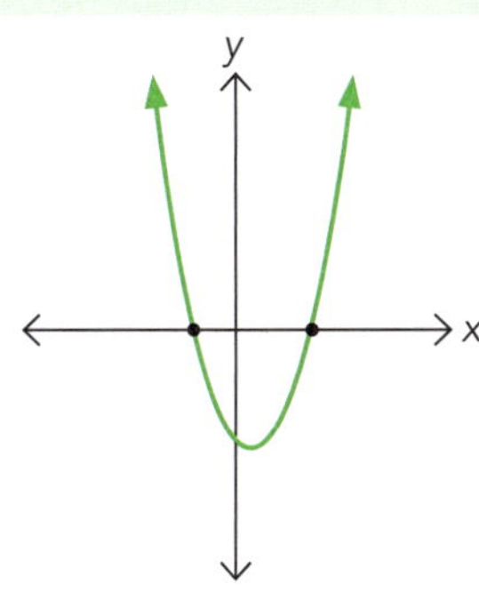

$\Delta = b^2 - 4ac > 0$

So $x = \dfrac{-b + \sqrt{+\text{ve number}}}{2a}$ or $\dfrac{-b - \sqrt{+\text{ve number}}}{2a}$

So there are **two** real distinct roots.

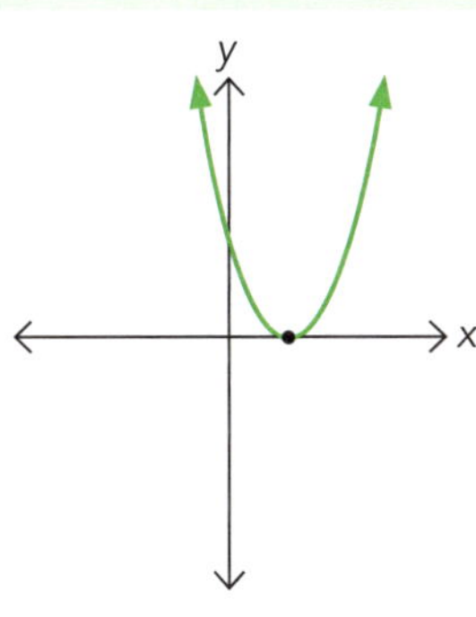

$\Delta = b^2 - 4ac = 0$

So $x = \dfrac{-b \pm \sqrt{0}}{2a}$ or $\dfrac{-b}{2a}$

So there is **one** real root.
(This could be described as two identical roots.)

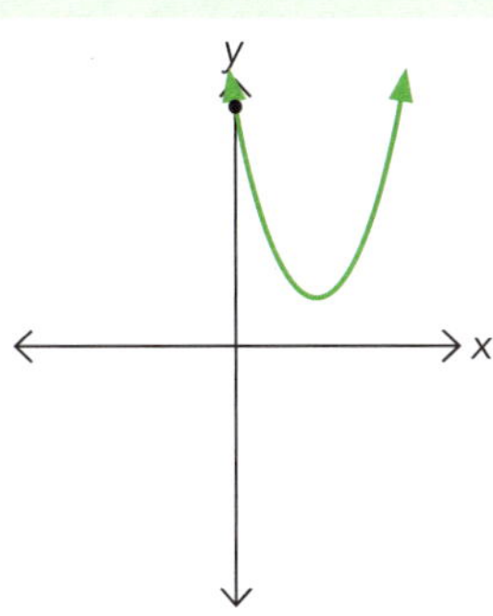

$\Delta = b^2 - 4ac < 0$

So $x = \dfrac{-b \pm \sqrt{-\text{ve number}}}{2a}$

But $\sqrt{\text{negative number}}$ does not exist.

So there are **no real** roots.

Example:

Calculate the value of the discriminant for $10x^2 - x - 3 = 0$, and use it to determine how many roots there are.

$$\begin{aligned} \Delta = b^2 - 4ac &= (-1)^2 - 4 \times 10 \times (-3) \\ &= 1 + 120 \\ &= +121 \end{aligned}$$

$\therefore$ therefore the discriminant is positive, so there are two real roots.

Using the discriminant to find values for a, b or c when given the number of roots

Don't forget:

- The square root of a number can be positive or negative.
- When you multiply or divide an inequality by a negative number, flip the sign.
- The roots occur when $y = 0$.
- Often you will need to convert the equation into the $ax^2 + bx + c$ format.

Examples:

1 For what values of p will the graph of $y = px^2 - 3x + 4$ not cut the x-axis?

'not cut the x-axis' $\Rightarrow$ no real roots $\Rightarrow b^2 - 4ac < 0$

$$b^2 - 4ac = (-3)^2 - 4 \times p \times 4 < 0$$
$$9 - 16p < 0$$
$$-16p < -9$$
$$p > 0.5625$$

Flip sign

So, $y = px^2 - 3x + 4$ will not cut the x-axis if $p > 0.5625$.

2 The equation $(3x - 1)(x + 5) = k$ has only one real solution. Find the value of k.

$$(3x - 1)(x + 5) = 3x^2 + 14x - 5 = k$$
$$3x^2 + 14x - 5 - k = 0$$

Convert to $ax^2 + bx + c$ format.

$$3x^2 + 14x - (5 + k) = 0$$
$$(14)^2 - (4 \times 3 \times -(5 + k)) = 0$$

One solution $\Rightarrow b^2 - 4ac = 0$.

$$196 + 60 + 12k = 0$$
$$12k = -256$$
$$k = -21.\dot{3}$$

So $(3x - 1)(x + 5) = k$ has only one real solution if $k = -21.\dot{3}$.

3 For what values of q will the equation $5x^2 + qx + 2 = 0$ have two real roots?

Two real roots $\Rightarrow b^2 - 4ac > 0$

$$q^2 - 4 \times 5 \times 2 > 0$$
$$q^2 - 40 > 0$$
$$q^2 > 40$$
$$\therefore \text{ either } q > 6.3246$$
$$\text{or } -q > 6.3246$$
$$q < -6.3246$$

So there will be two real roots if $q > 6.3246$ or $q < -6.3246$.

 ISBN: 9780170447058

Calculate the value of the discriminant and state the number of real roots.

1 $2x^2 + 5x - 12 = 0$

2 $9x^2 - 12x + 4 = 0$

3 $2x^2 - 7x + 10 = 0$

4 $4x^2 + 5x + 2 = 0$

Answer the following.

5 For what values of p will the graph of $y = px^2 - 7x + 4$ not cut the x-axis?

6 For what values of p will $y = 9x^2 + px + 11$ have two real distinct roots?

7 The equation $17 + x(16x - 11) - 13x = k$ has no real solutions. Find the value of k.

8 Find the possible values for d if real solutions exist for $9x^2 + (2 - 4d)x + 4d + 5 = 0$.

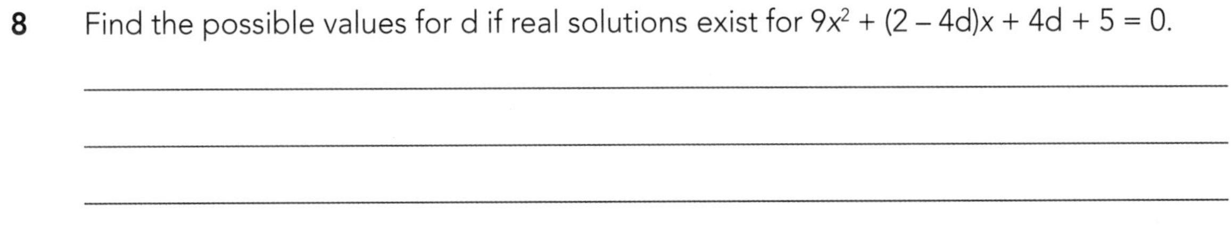

2 Circles

- A **locus** (plural: loci) is the set of all points that **share a property**. For example, the **locus** of the set of all the points that are equidistant from two defined points (**A** and **B**) is a **straight line**:

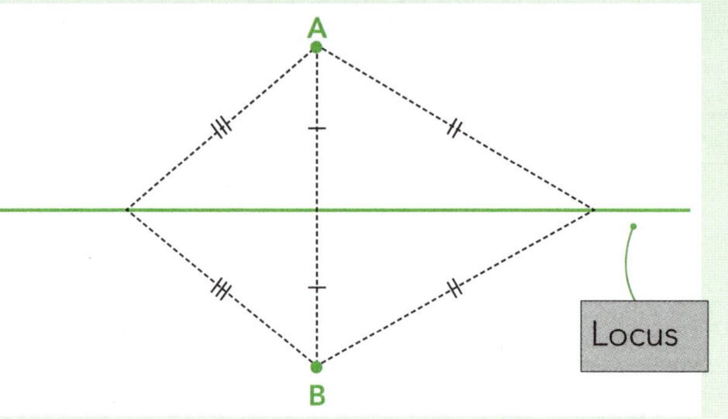

- A **circle** is the locus of all the points that are the same distance (r) from a point (**A**):

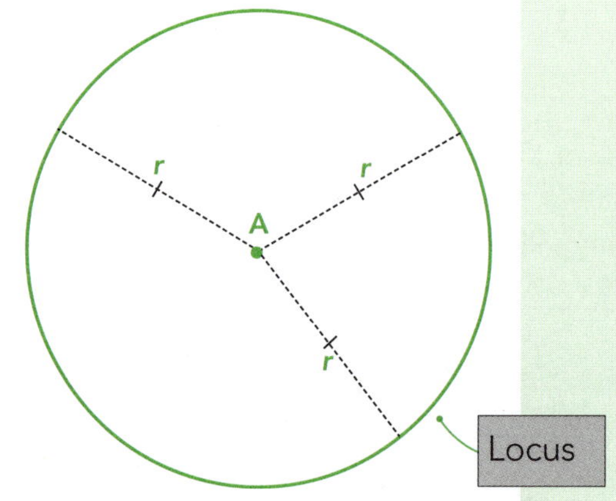

Circles with centres at the origin: have equations of the form $x^2 + y^2 = r^2$, where r represents the **radius** of the circle.

Examples:

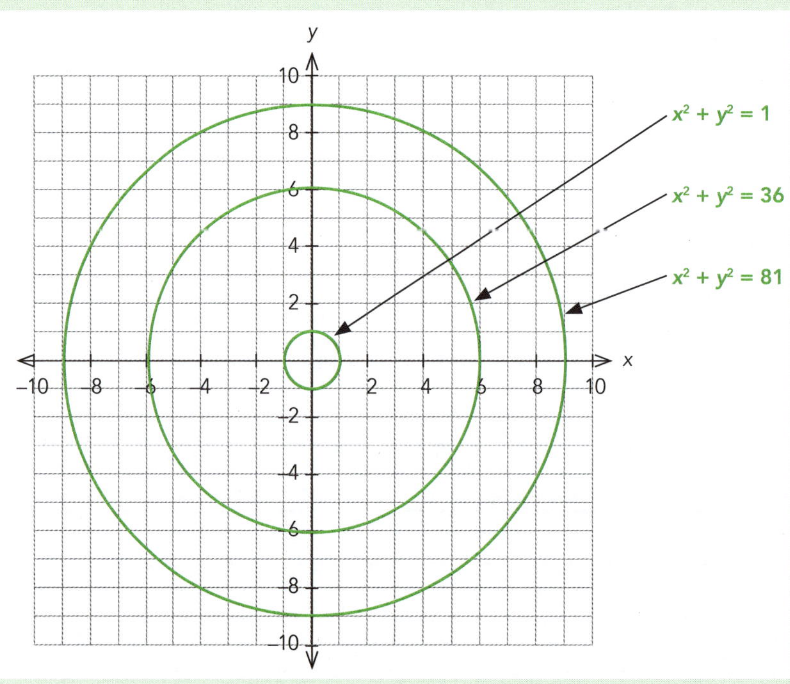

ISBN: 9780170447058

Circles with centres on the *x*-axis: horizontal translations: have equations of the form $(x + d)^2 + y^2 = r^2$, where *r* represents the radius of the circle, and **d** represents the **distance** the circle is translated.

Examples:

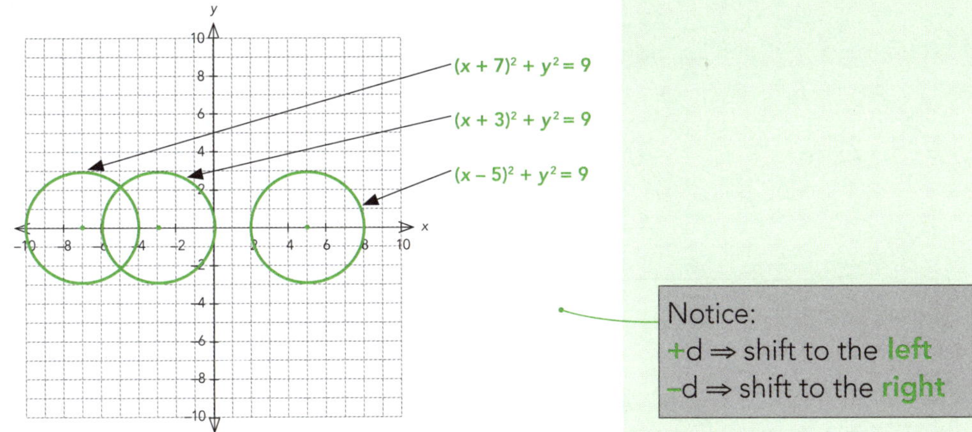

Circles with centres on the *y*-axis: vertical translations: have equations of the form $x^2 + (y + e)^2 = r^2$, where *r* represents the radius of the circle, and **e** represents the **distance** the circle is translated.

Examples:

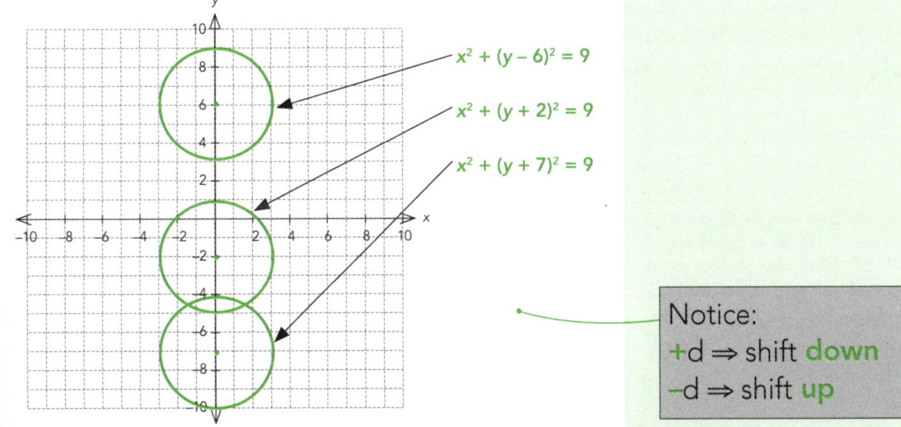

Combinations of vertical and horizontal translations: have equations of the form $(x \pm d)^2 + (y \pm e)^2 = r^2$.

Examples:

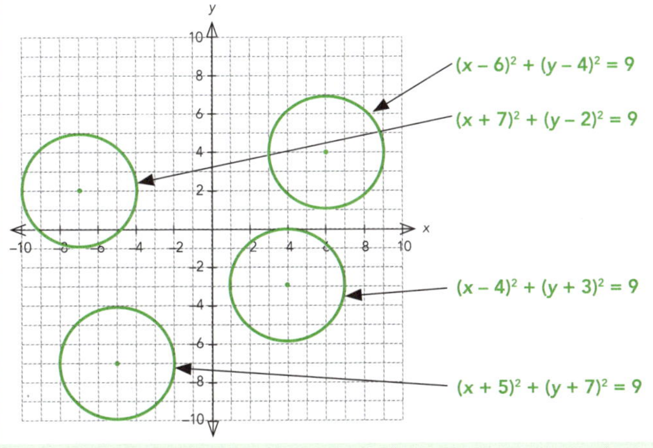

ISBN: 9780170447058

1 Match the circles drawn to their equations.

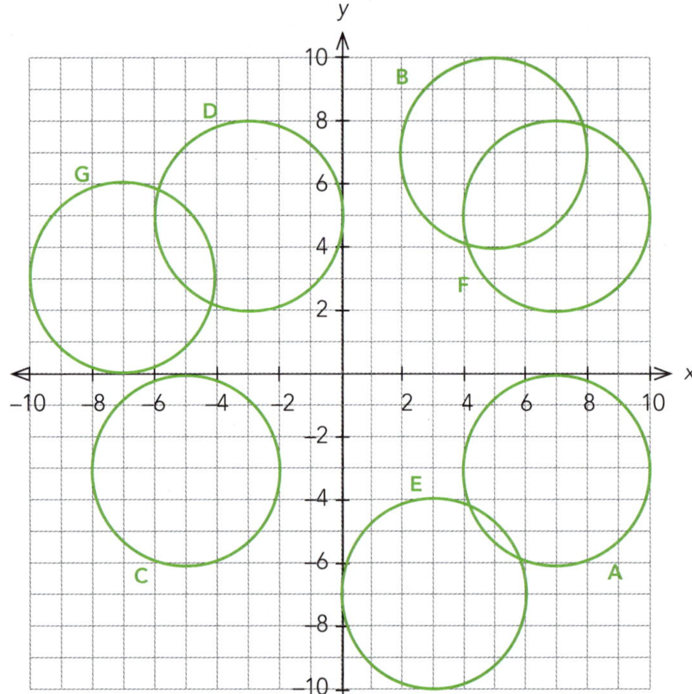

$(x + 7)^2 + (y - 3)^2 = 9$ ________

$(x + 5)^2 + (y + 3)^2 = 9$ ________

$(x - 3)^2 + (y + 7)^2 = 9$ ________

$(x - 7)^2 + (y + 3)^2 = 9$ ________

$(x - 5)^2 + (y - 7)^2 = 9$ ________

$(x + 3)^2 + (y - 5)^2 = 9$ ________

$(x - 7)^2 + (y - 5)^2 = 9$ ________

2 Match the circles drawn to their equations.

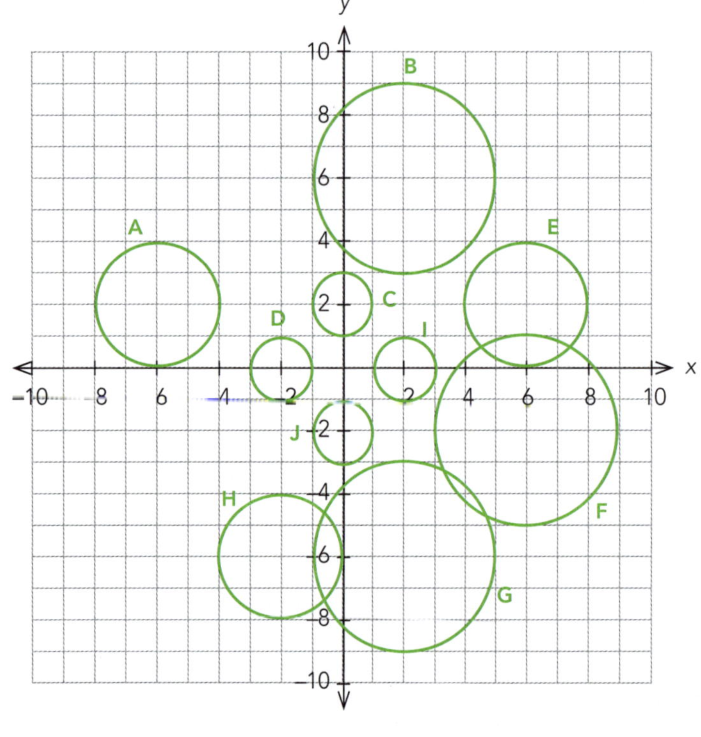

$(x + 6)^2 + (y - 2)^2 = 4$ ________

$x^2 + (y - 2)^2 = 1$ ________

$(x - 2)^2 + (y + 6)^2 = 9$ ________

$(x + 2)^2 + y^2 = 1$ ________

$(x - 2)^2 + (y - 6)^2 = 9$ ________

$(x + 2)^2 + (y + 6)^2 = 4$ ________

$x^2 + (y + 2)^2 = 1$ ________

$(x - 6)^2 + (y + 2)^2 = 9$ ________

$(x - 2)^2 + y^2 = 1$ ________

$(x - 6)^2 + (y - 2)^2 = 4$ ________

ISBN: 9780170447058

3 Trigonometry

Trigonometric functions

You will have learned about three trigonometric functions: **sin x** (sine)
cos x (cosine)
and **tan x** (tangent).

f(x) or y = sin x
In a right-angled triangle, $\sin x = \frac{\text{opposite}}{\text{hypotenuse}}$

f(x) or y = cos x
In a right-angled triangle, $\cos x = \frac{\text{adjacent}}{\text{hypotenuse}}$

f(x) or y = tan x
In a right-angled triangle, $\tan x = \frac{\text{opposite}}{\text{adjacent}}$

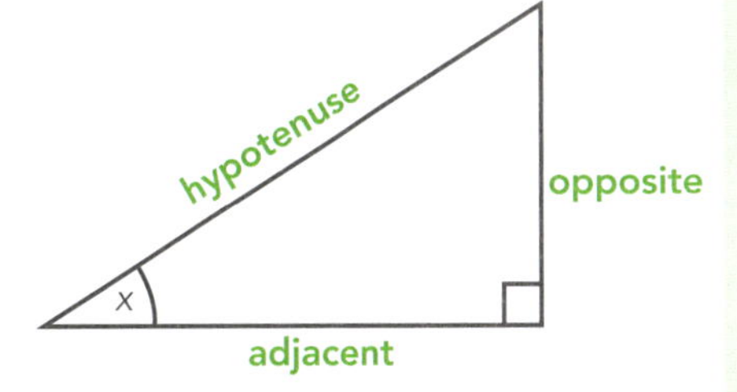

You should also be familiar with the behaviour of sine and cosine functions for angles less than 0° and greater than 90°.

f(x) = sin x:

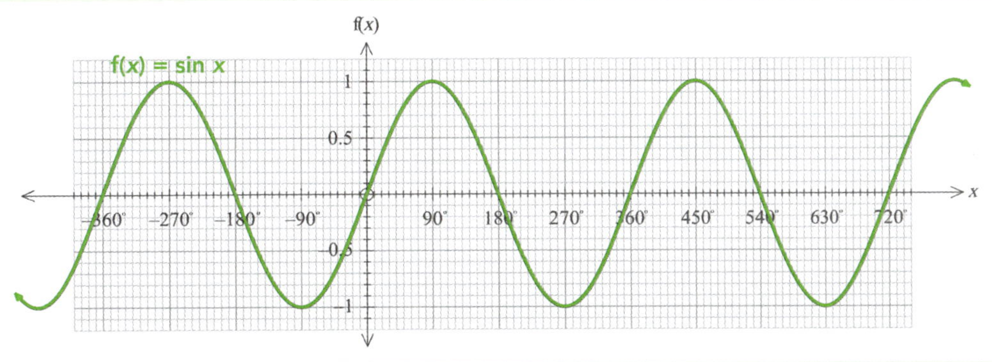

f(x) = cos x:

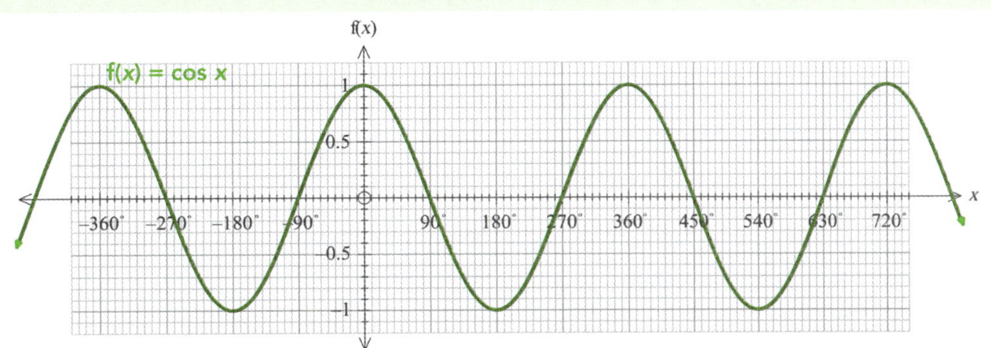

These are known as **periodic** functions.
Instead of measuring angles in degrees, you will be measuring them in **radians**.

Radians

- One radian = the angle formed in a sector with an arc length that is the same as the radius.
- One radian is approximately **57.3°**.

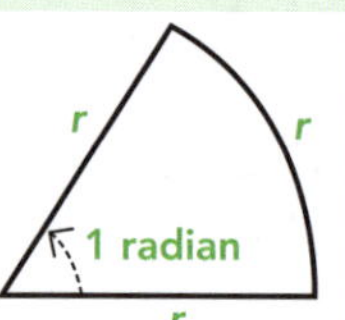

- The length of the radius doesn't make any difference to the angle size:

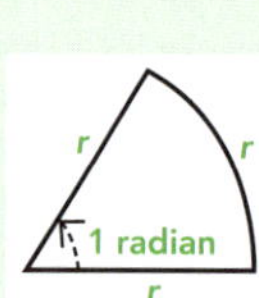

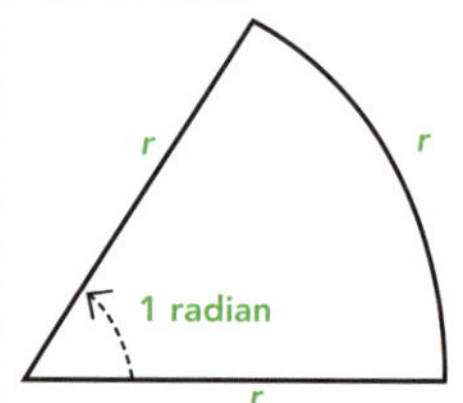

- Make sure you know how to switch your calculator between degrees and radians.

To convert between radians and degrees:

$$\frac{1 \text{ radian}}{360°} = \frac{r}{\text{circumference of the circle}}$$

$$= \frac{r}{2\pi r}$$

$$= \frac{1}{2\pi}$$

$\therefore 2\pi$ radians = 360°

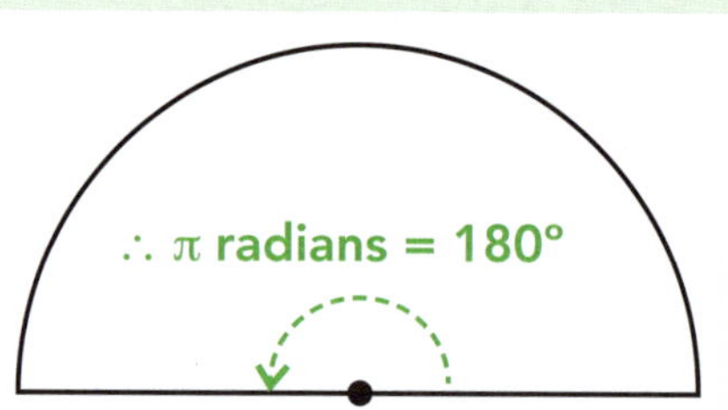

Example: Convert 45° to radians.

$180° = \pi$ radians

$$\therefore 45° = \pi \times \frac{45°}{180°}$$

$$= \pi \times \frac{1}{4}$$

$$= 0.7854 \text{ or } \frac{\pi}{4}$$

It is often convenient to write angles in terms of π.

Example: Convert $\frac{2\pi}{3}$ (radians) to degrees.

Whenever you see an angle which includes π, that angle will be in radians.

π radians = 180°

$$\therefore \frac{2}{3}\pi = \frac{2}{3} \times 180°$$

$$= 120°$$

 ISBN: 9780170447058

Convert the following angles from degrees to radians.

1 180° = ______________________

2 90° = ______________________

3 30° = ______________________

4 150° = ______________________

Convert the following angles from radians to degrees.

5 $\frac{\pi}{3}$ = ______________________

6 $\frac{\pi}{4}$ = ______________________

7 $\frac{3\pi}{4}$ = ______________________

8 $\frac{2\pi}{3}$ = ______________________

Complete the following diagrams, leaving π in your answers.

9

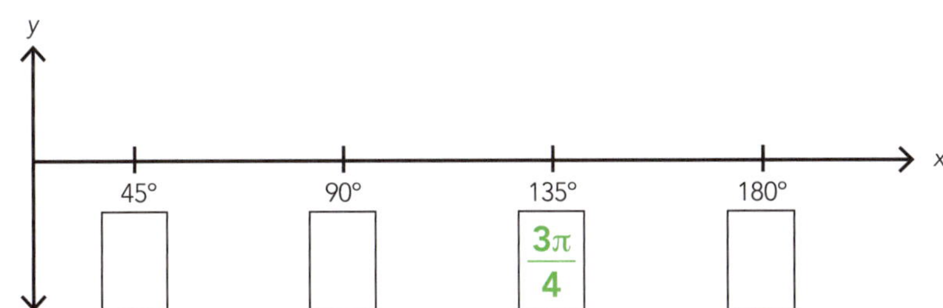

10

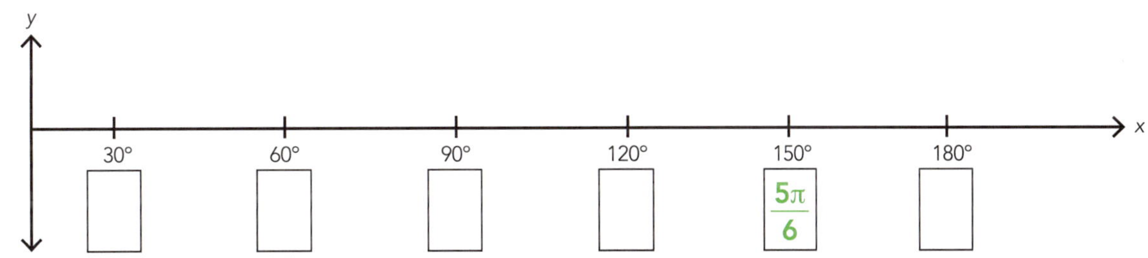

Exact values for trigonometric functions: special triangles

- Trigonometric functions for **most angles do not have exact values**.
- Your calculator gives you these values to many decimal places, and it is usual to **round** these to **4 decimal places**.
  For example: $\cos 0.6 = 0.825335614\ldots = 0.8253$ (4 dp)
- Using **special triangles**, it is possible to write **exact values** for some angles.
- **Exact** values must be able to be written as fractions made up of **whole numbers** and/or **surds** (roots of whole numbers).

For $\frac{\pi}{4}$ (or 45°), use an isosceles right-angled triangle with equal sides 1 unit long:

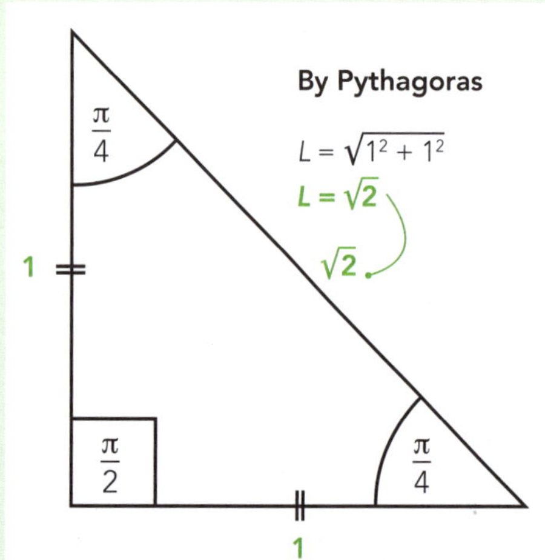

Using this triangle:

$\sin \frac{\pi}{4} = \frac{1}{\sqrt{2}}$

$\cos \frac{\pi}{4} = \frac{1}{\sqrt{2}}$

$\tan \frac{\pi}{4} = 1$

For $\frac{\pi}{6}$ and $\frac{\pi}{3}$ (or 30° and 60°), use half an equilateral triangle with sides 2 units long:

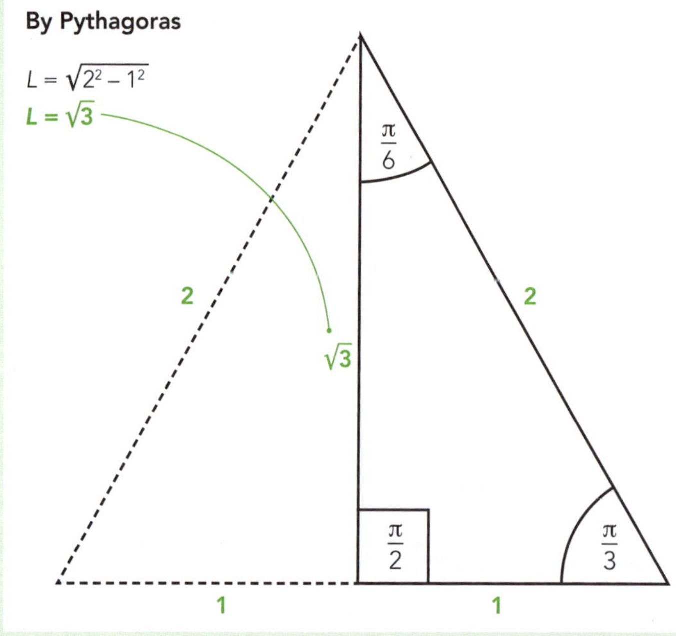

Using this triangle:

$\sin \frac{\pi}{6} = \frac{1}{2}$ $\quad$ $\sin \frac{\pi}{3} = \frac{\sqrt{3}}{2}$

$\cos \frac{\pi}{6} = \frac{\sqrt{3}}{2}$ $\quad$ $\cos \frac{\pi}{3} = \frac{1}{2}$

$\tan \frac{\pi}{6} = \frac{1}{\sqrt{3}}$ $\quad$ $\tan \frac{\pi}{3} = \sqrt{3}$

ISBN: 9780170447058

Another way of viewing trigonometric functions

- It is often convenient to view trigonometric functions in terms of quadrants.
- When dealing with complex numbers, the **principal value of θ** is restricted to $-\pi < \theta \leq \pi$.
- Angles in the **third and fourth quadrants** take **negative values** for θ.

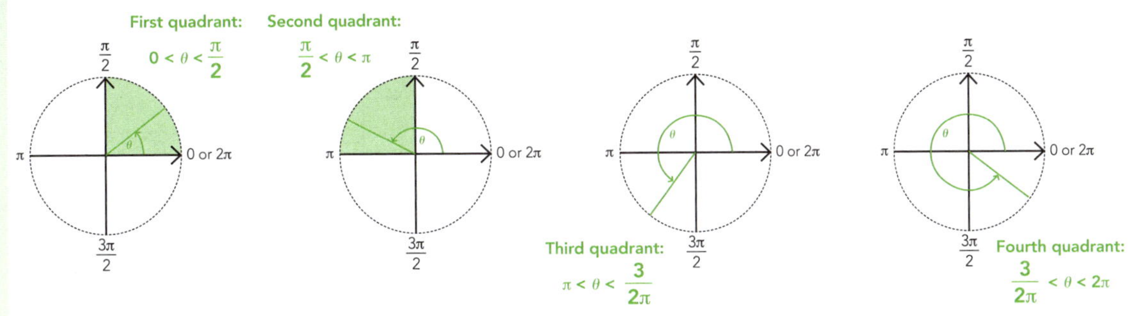

- It is also useful to be aware of the **signs** of each trigonometric function in each quadrant.

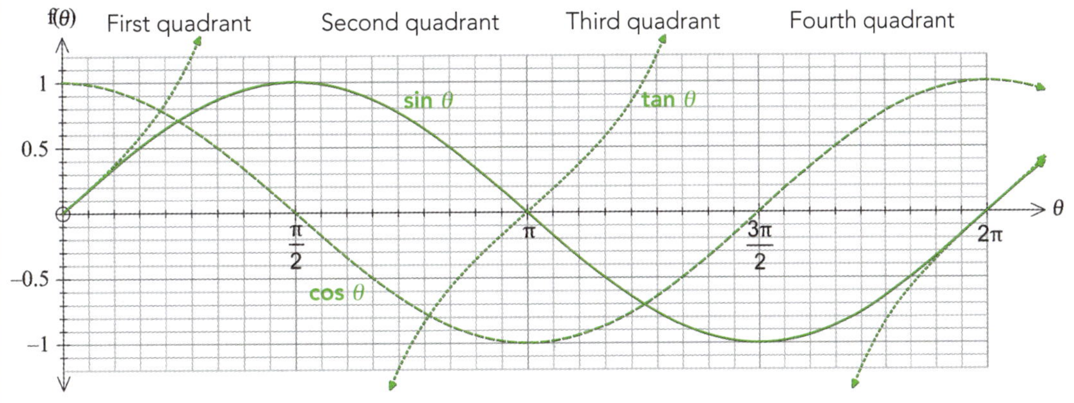

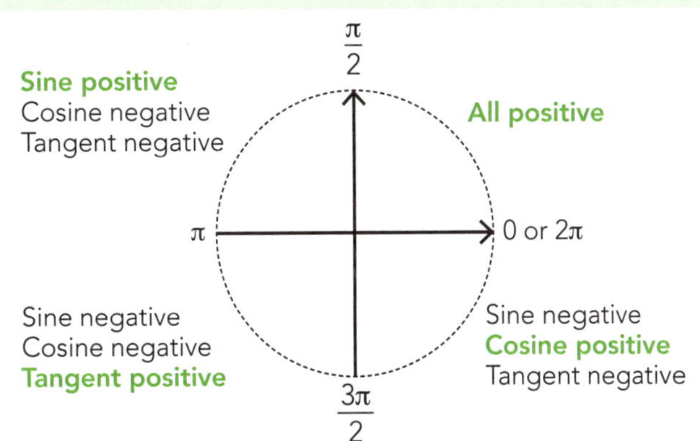

- This approach makes it easy to find other functions when told the value of just one function.

Examples:

1 If $\sin\theta = \frac{1}{2}$, find all possible values for θ.

sin θ is positive ⇒ θ must be in the first and second quadrants.

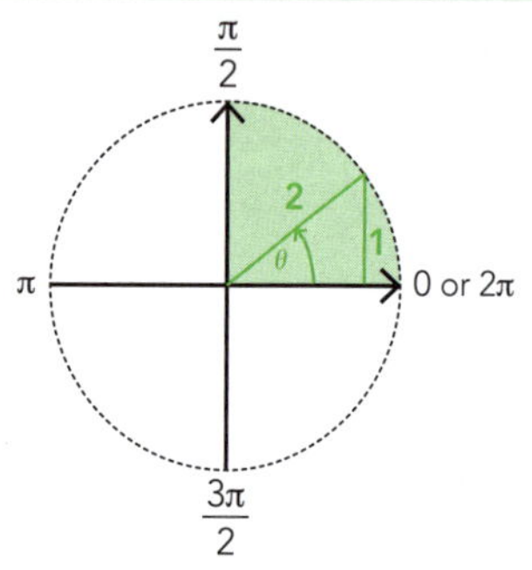

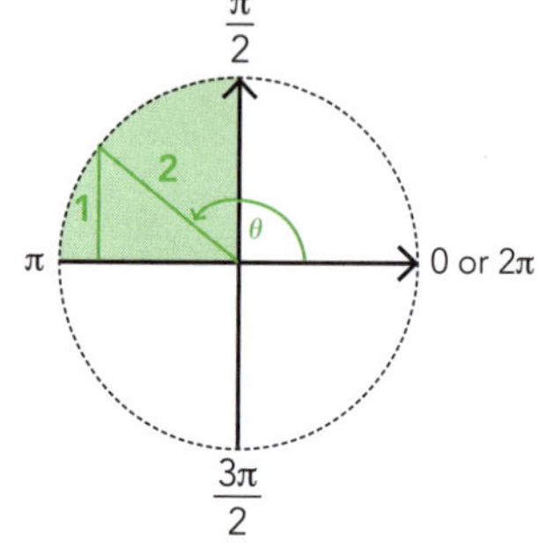

$$\theta = \sin^{-1}\frac{1}{2}$$

$$\therefore\ \theta = \frac{\pi}{6} \text{ or } 30° \qquad \text{or} \qquad \theta = \pi - \frac{\pi}{6} = \frac{5\pi}{6} \text{ or } 150°$$

Note: You could also do this using a graph:

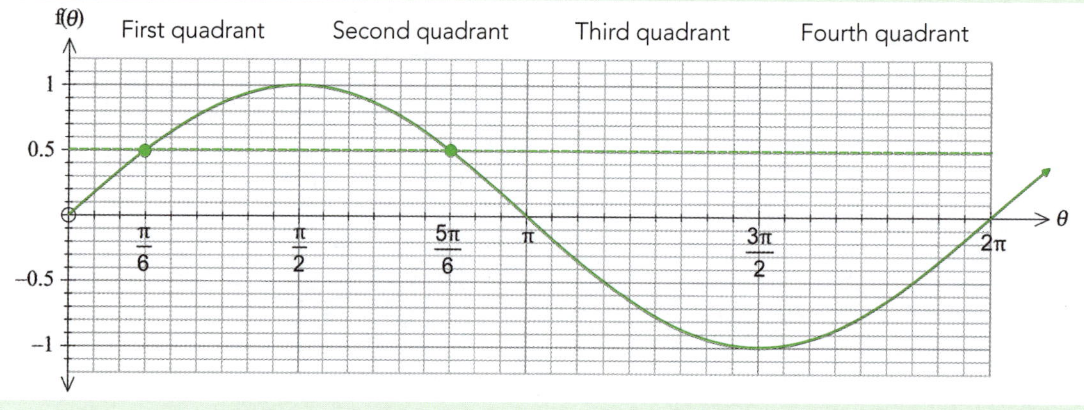

2 If $\tan\theta = -1$, find the values of θ.

tan θ is negative ⇒ θ must be in the second and fourth quadrants.

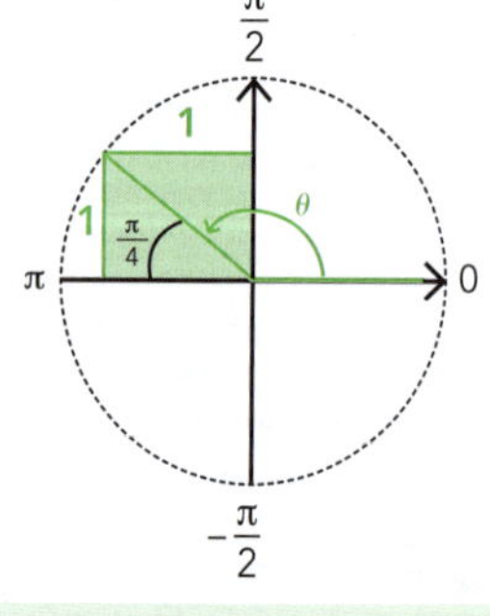

$$\theta = \tan^{-1}(-1)$$

$$\therefore\ \theta = \pi - \frac{\pi}{4} = \frac{3\pi}{4} \text{ or } 135° \quad \text{or} \qquad \theta = -\frac{\pi}{4} \text{ or } -45°$$

ISBN: 9780170447058

Putting it together

Use your calculator to find the following sines, cosines and tangents of angles. Make sure you switch your calculator to **radians** and use **brackets** when there are fractions.

1 $\sin \pi =$ ____________________

2 $\cos 2\pi =$ ____________________

3 $\tan \frac{\pi}{4} =$ ____________________

4 $\sin \frac{3\pi}{2} =$ ____________________

5 $\cos \frac{\pi}{3} =$ ____________________

6 $\tan (-\frac{\pi}{4}) =$ ____________________

7 $\cos \frac{\pi}{4} =$ ____________________

8 $\sin \frac{4\pi}{3} =$ ____________________

Give your answers to the following in radians and in degrees.

9 If $\cos \theta = \frac{1}{2}$ and $0 < \theta < 2\pi$, find all possible values for θ.

____

____

____

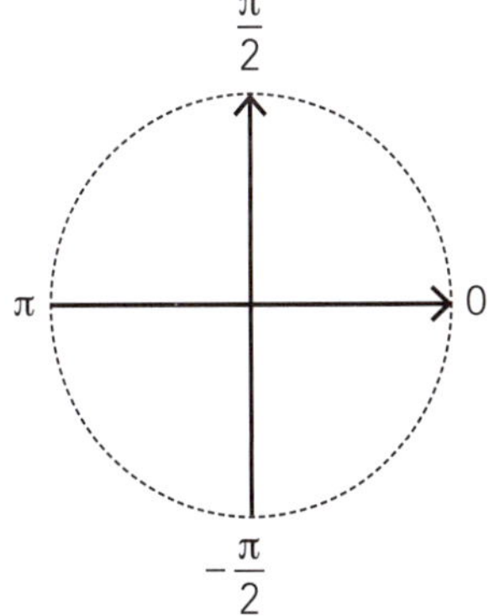

10 If $\cos \theta = -\frac{1}{2}$ and $0 < \theta < 2\pi$, find all possible values for θ.

____

____

____

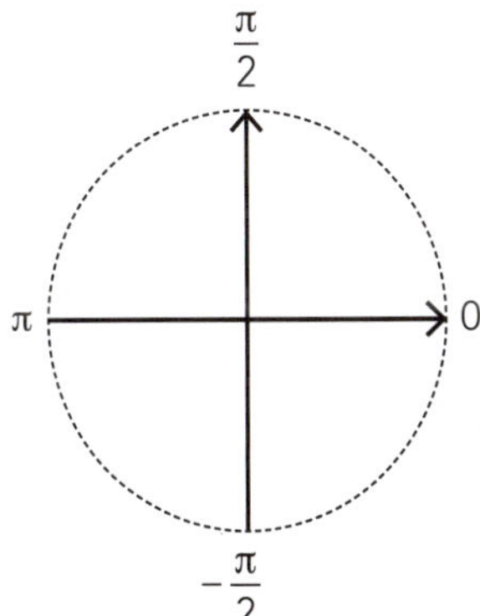

ISBN: 9780170447058

11 If $\tan \theta = 1$ and $0 < \theta < 2\pi$, find all possible values for θ.

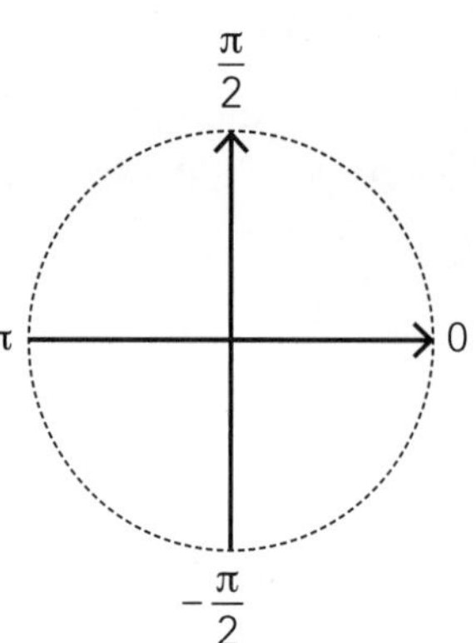

12 If $\cos \theta = \frac{1}{2}$ and $0 < \theta < 2\pi$, find all possible values for θ.

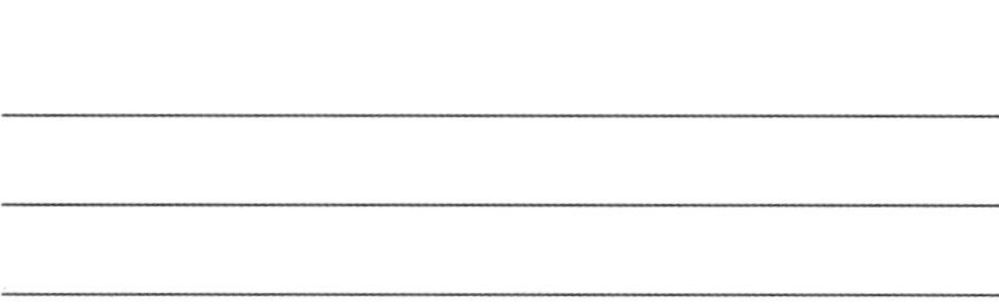

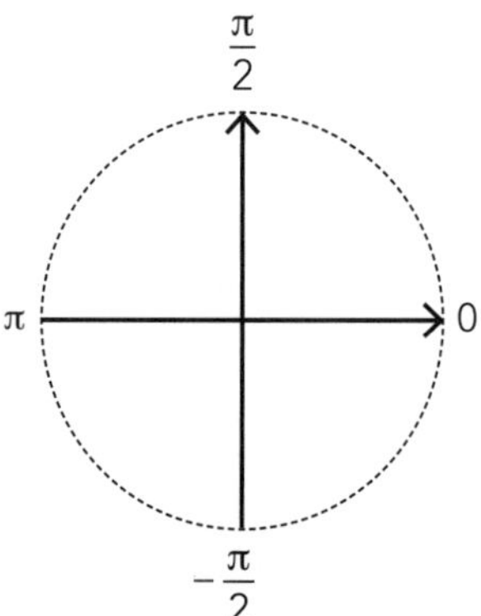

13 If $\cos \theta = -\frac{\sqrt{3}}{2}$ and $0 < \theta < 2\pi$, find all possible values for θ.

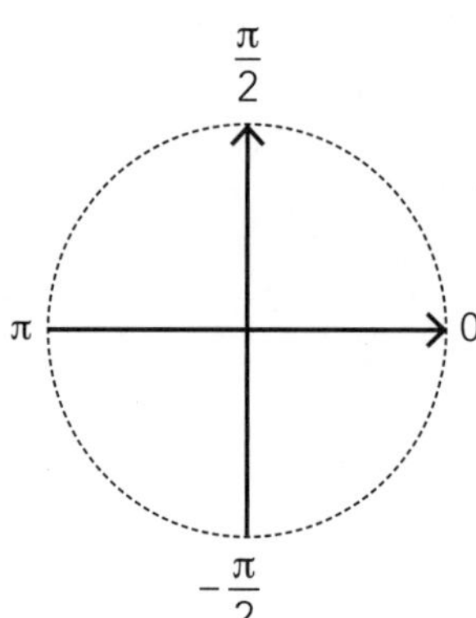

14 If $\sin \theta = 1$ and $0 < \theta < 2\pi$, find all possible values for θ.

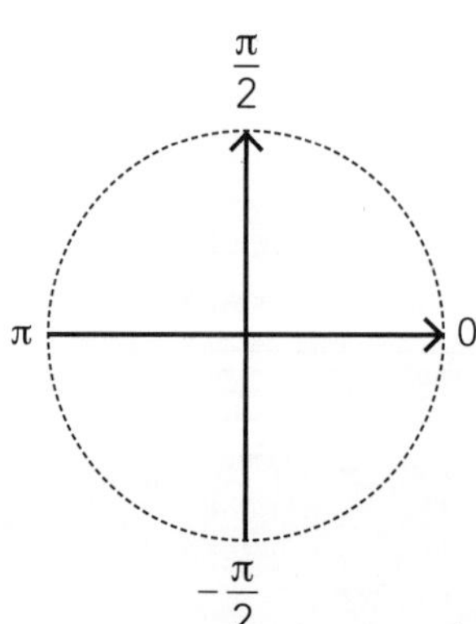

ISBN: 9780170447058

Groups of numbers

- For this standard, it will be very helpful if you understand the relationships between the different groups of numbers.

Real numbers (R)

Rational numbers (Q)
Can be expressed as a fraction.
Includes terminating and recurring decimals.
e.g. $\frac{3}{7}$, 1.49, $0.8\dot{1}\dot{3}$

Integers (*I*)
... –4, –3, –2, –1

Whole numbers (*W*)
0

Natural or counting numbers (*N*)
1, 2, 3, 4, ...

Irrational numbers (Q')
Non-terminating decimals including surds.
e.g. $\sqrt{2}$, π, e

Unreal or imaginary numbers

Expressed in terms of $i = \sqrt{-1}$
e.g. 13i, –4i

You will get to know all about these a little later in this book.

a (real numbers) — *b* (unreal or imaginary numbers)

Complex numbers (Z)

Made up of:
1 a **real** part (*a*)
2 an **imaginary** part (*b*)

$z = a + bi$

where
a = *Re*(z) and b = *Im*(z)

Re(z) means the **real** part of z.

Im(z) means the **imaginary** part of z.

ISBN: 9780170447058

Surds

- A surd is an irrational number that can't be simplified by removal of the root sign.
- Numbers that **are** surds: $\sqrt{2}$, $\sqrt[3]{10}$, $\sqrt[4]{32}$, etc.
- Numbers that **are not** surds: $\sqrt{9}$ (= 3), $\sqrt[3]{64}$ (= 4), $\sqrt[5]{32}$ (= 2).
- Remember that a $\sqrt{\ }$ sign means the **square** root of or $\sqrt[2]{\ }$.

Rules for surds

Hint: It can often be useful to use the Year 9 method of reducing a number to prime factors.

Example:

2	540
2	270
3	135
3	45
3	15
5	5
	1

$\therefore\ 540 = 2^2 \times 3^2 \times 3 \times 5$

This tells you that 540 is not a power of a number, so $\sqrt{540}$ is a surd, but also that it can be written as

$2 \times 3\sqrt{3 \times 5} = 6\sqrt{15}$.

Surds and powers $\sqrt[n]{a^n} = a$

1 $\sqrt{49} = \sqrt{7^2}$
$= 7$

2 $\sqrt[6]{64} = \sqrt[6]{2^6}$
$= 2$

3 $\sqrt{p^4q^6} = p^2q^3$

Remember, when finding a square root, you halve the powers.

4 $\sqrt{162p^3q^4} = \sqrt{9^2 \times 2 \times p^2 \times p \times q^4}$
$= 9pq^2\sqrt{2p}$

When simplifying a square root, it is useful to write as much as possible with **even** numbers as powers.

Multiplying surds $\sqrt{ab} = \sqrt{a} \times \sqrt{b}$

1 $\sqrt{63} = \sqrt{9} \times \sqrt{7}$
$= \sqrt{3^2} \times \sqrt{7}$
$= 3\sqrt{7}$

2 $\sqrt{180} - \sqrt{20} = \sqrt{36} \times \sqrt{5} - \sqrt{4} \times \sqrt{5}$
$= 6\sqrt{5} - 2\sqrt{5}$
$= 4\sqrt{5}$

3 $3\sqrt{5} \times 2\sqrt{12} = 6\sqrt{60}$
$= 6\sqrt{2^2 \times 15}$
$= 12\sqrt{15}$

4 $\sqrt{32p^3q^2} - p\sqrt{18pq^2} = 4pq\sqrt{2p} - 3pq\sqrt{2p}$
$= pq\sqrt{2p}$

ISBN: 9780170447058

Dividing surds

$$\frac{\sqrt{a}}{\sqrt{b}} = \sqrt{\frac{a}{b}}$$

1 $\frac{\sqrt{15}}{\sqrt{27}} = \sqrt{\frac{15}{27}}$

$= \sqrt{\frac{5}{9}}$

$= \frac{\sqrt{5}}{3}$

2 $\sqrt{\frac{54}{225}} \times \sqrt{6} = \sqrt{\frac{3^2 \times 6}{3^2 \times 5^2}} \times \sqrt{6}$

$= \frac{\sqrt{6} \times \sqrt{6}}{5}$

$= \frac{6}{5}$

3 $\frac{p\sqrt{2q}}{q\sqrt{p}} \div \frac{\sqrt{2p}}{\sqrt{q^3}} = \frac{p\sqrt{2q}}{q\sqrt{p}} \times \frac{\sqrt{q^3}}{\sqrt{2p}}$

$= \frac{p\sqrt{2q^4}}{q\sqrt{2p^2}}$

$= \frac{pq^2\sqrt{2}}{pq\sqrt{2}}$

$= q$

Simplify the following.

1 $\sqrt{81}$

2 $\sqrt[3]{27}$

3 $\sqrt[6]{64}$

4 $\sqrt[4]{256}$

5 $\sqrt{x^2y^6}$

6 $\sqrt[3]{8x^6y^3z^{12}}$

7 $\sqrt{75p^5q^{10}r}$

8 $\sqrt{\frac{80q^6}{25p^2}}$

Write each of the following in the form $a\sqrt{b}$.

9 $\sqrt{12}$

10 $\sqrt{45}$

11 $\sqrt{252}$

12 $\sqrt{432}$

13 $\sqrt{49p^2q}$

14 $\sqrt{\frac{xy^2}{4}}$

ISBN: 9780170447058

Simplify the following.

15 $\sqrt{48} + \sqrt{75}$

16 $\sqrt{147} + \sqrt{108} - \sqrt{192}$

17 $4\sqrt{9} \times 3\sqrt{8}$

18 $\dfrac{\sqrt{28}}{\sqrt{36}}$

19 $5a\sqrt{6b^3} \times \sqrt{4a^2b}$

20 $b\sqrt{75ab^3} - \sqrt{27ab^5}$

21 $\dfrac{\sqrt{44}}{\sqrt{108}}$

22 $\dfrac{2p\sqrt{21p}}{\sqrt{12qp^2}}$

23 $\dfrac{7\sqrt{6}}{11\sqrt{15}}$

24 $\dfrac{3\sqrt{10}}{4\sqrt{21}} \div \dfrac{5\sqrt{24}}{12\sqrt{12}}$

ISBN: 9780170447058

Simplifying expressions containing surds

- This requires the combination of the usual rules for expansion with the rules for surds.

Examples: Expand and simplify the following.

1 $\sqrt{2}(\sqrt{5}-3) = \sqrt{2}\sqrt{5} - 3\sqrt{2}$
$= \sqrt{10} - 3\sqrt{2}$

2 $(\sqrt{x} - 2\sqrt{y})(y - 3\sqrt{xy}) = y\sqrt{x} - 3x\sqrt{y} - 2y\sqrt{y} + 6y\sqrt{x}$
$= 7y\sqrt{x} - \sqrt{y}(3x + 2y)$

Expand and simplify the following.

1 $\sqrt{7}(\sqrt{2} + \sqrt{3})$

2 $\sqrt{3}(5\sqrt{2} - 2\sqrt{3})$

3 $(\sqrt{3} - 5)(\sqrt{2} + 6)$

4 $(\sqrt{5} - 1)(\sqrt{5} + 1)$

5 $(\sqrt{7} - 1)^2$

6 $\sqrt{x}\,(3\sqrt{x} - \sqrt{xy})$

7 $(\sqrt{p} - \sqrt{q})^2$

8 $(2\sqrt{x} + \sqrt{3y})^2$

9 $(2\sqrt{x} + 3)(5 - 7\sqrt{x})$

10 $(\sqrt{x} - 2)^3$

ISBN: 9780170447058

Rationalising and more simplifying of expressions containing surds

- Rationalising a surd means turning it into an expression that does **not** have a surd in the **denominator**.
- **You will need to do this often!**

Examples: Rationalise the following.

1 $\frac{5}{\sqrt{3}} = \frac{5}{\sqrt{3}} \times \frac{\sqrt{3}}{\sqrt{3}}$

Remember that multiplying by $\frac{a}{a}$ is the same as multiplying by 1.

$= \frac{5\sqrt{3}}{3}$

No surd in the denominator.

2 $\frac{7p}{\sqrt{pq}} = \frac{7p}{\sqrt{pq}} \times \frac{\sqrt{pq}}{\sqrt{pq}}$

$= \frac{7p\sqrt{pq}}{pq}$

$= \frac{7\sqrt{pq}}{q}$

3 $\frac{1}{3\sqrt{pq}} = \frac{1}{3\sqrt{pq}} \times \frac{\sqrt{pq}}{\sqrt{pq}}$

$= \frac{\sqrt{pq}}{3pq}$

Rationalise and simplify the following.

1 $\frac{3}{\sqrt{6}}$

2 $\frac{8x}{\sqrt{x}}$

3 $\frac{5}{2\sqrt{p}}$

4 $\frac{6x}{\sqrt{2xy}}$

5 $\frac{3}{\sqrt{a-b}}$

6 $\frac{1}{3\sqrt{2x-1}}$

ISBN: 9780170447058

Useful trick:

- Remember that $(a + b)(a - b) = a^2 - b^2$.

Notice that both of these terms are squares.

- Because the terms on the right are squares, we can use this to rationalise (get rid of) surds on the bottom of a fraction.
- $(\sqrt{a} + \sqrt{b})$ and $(\sqrt{a} - \sqrt{b})$ are known as the **conjugate surds**.

$$(\sqrt{a} + \sqrt{b})(\sqrt{a} - \sqrt{b}) = a - b$$

No surds on the right side.

Examples:

1 $\dfrac{7}{\sqrt{2}+3} = \dfrac{7}{\sqrt{2}+3} \times \dfrac{\sqrt{2}-3}{\sqrt{2}-3}$

$= \dfrac{7(\sqrt{2}-3)}{2-9}$

$= \dfrac{7(\sqrt{2}-3)}{-7}$

$= 3 - \sqrt{2}$

2 $\dfrac{3}{2-\sqrt{p}} = \dfrac{3}{2-\sqrt{p}} \times \dfrac{2+\sqrt{p}}{2+\sqrt{p}}$

$= \dfrac{3(2+\sqrt{p})}{4-p}$

3 $\dfrac{7}{1-\sqrt{5}} - \dfrac{5}{1+\sqrt{5}} = \dfrac{7(1+\sqrt{5}) - 5(1-\sqrt{5})}{(1-\sqrt{5})(1+\sqrt{5})}$

$= \dfrac{7 + 7\sqrt{5} - 5 + 5\sqrt{5}}{(1-\sqrt{5})(1+\sqrt{5})}$

$= \dfrac{2 + 12\sqrt{5}}{-4}$

$= -\dfrac{1}{2} - 3\sqrt{5}$

Write each expression in the form $a \pm b\sqrt{c}$.

7 $\dfrac{2}{\sqrt{3}+1}$

8 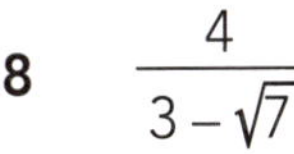
$\dfrac{4}{3-\sqrt{7}}$

ISBN: 9780170447058

9 $\frac{36}{5+\sqrt{7}}$

10 $\frac{88}{7-\sqrt{5}}$

11 $\frac{\sqrt{12}}{1-\sqrt{3}}$

12 $\frac{1+\sqrt{7}}{\sqrt{7}-2}$

Rationalise the following.

13 $\frac{1}{1-\sqrt{a}}$

14 $\frac{2\sqrt{p}}{3\sqrt{p}+1}$

Add or subtract the following, writing your answer as a rational expression.

15 $\frac{5}{2-\sqrt{3}}-\frac{1}{2+\sqrt{3}}$

16 $\frac{2}{\sqrt{5}+1}+\frac{3}{\sqrt{5}-1}$

17 $\frac{1}{\sqrt{x}+1}+\frac{2}{\sqrt{x}-1}$

18 $\frac{5}{2-\sqrt{y}}+3$

ISBN: 9780170447058

Equations involving surds

Step 1: If possible, rearrange the equation so there is a **maximum** of one surd term on each side.
Step 2: **Square** both sides.
Step 3: Follow the usual rules for solving equations.
Step 4: **Important:** The process of squaring may add extra roots, so **you must check your answers**.

Example:

Let $x = -1$
Square both sides: $x^2 = 1$
Square root: $x = \pm 1$

Note the extra solution!

Examples: Solve the following.

1

$$3\sqrt{x+2} + x = 2$$
$$3\sqrt{x+2} = 2 - x$$
Rearrange so the surd expression is on its own on one side.
$$9(x+2) = (2-x)^2$$
Square both sides.
$$9x + 18 = 4 - 4x + x^2$$
$$x^2 - 13x - 14 = 0$$
Rearrange into quadratic form.
$$(x+1)(x-14) = 0$$
$$\therefore \quad x = -1 \text{ or } x = 14$$

Check: $x = -1 \Rightarrow 3\sqrt{-1+2} + -1 = 2$ ✓
$x = 14 \Rightarrow 3\sqrt{14+2} + 14 = 26 \neq 2$ ✗

$\therefore$ **Solution is $x = -1$.**

2

$$\sqrt{5x+6} - 2 = \sqrt{2x}$$
$$\sqrt{5x+6} = \sqrt{2x} + 2$$
Rearrange so that the 'messier' surd is on its own on one side.
$$5x + 6 = 2x + 4\sqrt{2x} + 4$$
Square both sides.
$$3x + 2 = 4\sqrt{2x}$$
Rearrange so that the surd is on its own on one side.
$$9x^2 + 12x + 4 = 32x$$
Square both sides again.
$$9x^2 - 20x + 4 = 0$$
$$(x-2)(9x-2) = 0$$
$$x = 2 \text{ or } 0.\dot{2}$$

Check: $x = 2 \Rightarrow \sqrt{10+6} - 2 = \sqrt{4}$ ✓
$x = 0.\dot{2} \Rightarrow \sqrt{1.\dot{1}+6} - 2 = \sqrt{0.\dot{4}}$ ✓

$\therefore$ **Solution is $x = 2$ or $0.2\dot{0}$.**

ISBN: 9780170447058

Solve the following.

1 $3 + \sqrt{x + 3} = x$

2 $\sqrt{3x + 22} = x + 4$

3 $\sqrt{6 - 2x} + 3 = x$

4 $x - 2\sqrt{x + 6} = 2$

5 $\sqrt{7x + 1} = 1 + \sqrt{5x}$

6 $\sqrt{3x + 4} - \sqrt{2x + 2} = 1$

ISBN: 9780170447058

Sometimes you may be asked to give the solution in terms of another variable.

- The same rules apply, but with these, you do not need to check your answer.

Examples: Solve the following for x, giving your answer in terms of k.

1

$$\sqrt{x} + 3 = \sqrt{x + 2k}$$

$$x + 6\sqrt{x} + 9 = x + 2k$$ — Square both sides.

$$6\sqrt{x} = 2k - 9$$ — Rearrange so the surd expression is on its own on one side.

$$36x = (2k - 9)^2$$ — Square both sides again to get rid of the surd.

$$x = \frac{(2k - 9)^2}{36}$$

2

$$7\sqrt{x} - 2k\sqrt{x - 3} = 0$$

$$7\sqrt{x} = 2k\sqrt{x - 3}$$ — Rearrange so that there is one surd term on each side.

$$49x = 4k^2(x - 3)$$ — Square both sides.

$$49x = 4k^2x - 12k^2$$

$$4k^2x - 49x = 12k^2$$ — Rearrange so x terms are on one side.

$$x(4k^2 - 49) = 12k^2$$ — Factorise.

$$x = \frac{12k^2}{4k^2 - 49}$$

Solve the following equations for x, giving your answer in terms of k.

7 $\sqrt{x - k} = 4 + \sqrt{x}$

8 $\sqrt{x - 5k} = 11 - \sqrt{x}$

9 $\sqrt{x} = \sqrt{2k + x} + 3$

10 $\sqrt{7k + x} - \sqrt{x} = -1$

11 $\sqrt{x} = k\sqrt{x + 1}$

12 $k\sqrt{x + 4} - 3\sqrt{x} = 0$

13 $3\sqrt{7 + x} = 4k\sqrt{x}$

14 $k\sqrt{2x - 1} - \sqrt{5 + x} = 0$

ISBN: 9780170447058

Polynomials

Completing the square

- You know how to solve quadratic equations by factorising and by using the quadratic formula.
- Completing the square is a third method for solving quadratic equations.
- It also provides a method for proving that a quadratic expression is always positive.

Perfect squares:

- A quadratic expression in the form $(x \pm a)^2$ is known as a perfect square.
- Finding the square root of this is easy: $\sqrt{(x \pm a)^2} = x \pm a$.
- If we can write a quadratic equation in the form $(x \pm a)^2 = b$, then $x \pm a = \pm\sqrt{b}$.

Examples:

1 Solve: $x^2 + 10x + 16 = 0$

Step 1: Using the coefficient of the x term, create an extra constant in order to produce a perfect square:

$$x^2 + \mathbf{10}x + \mathbf{25}$$

$$\left(\frac{10}{2}\right)^2 = \mathbf{25}$$

This equals $(x + 5)^2$, which is a perfect square.

Step 2: Because you have added an extra **25**, you need to subtract 25:

$$(x^2 + 10x + 25) - 25 + 16 = 0$$

Step 3: Rewrite the equation as a perfect square and a single constant:

$$(x + 5)^2 - 25 + 16 = 0$$

$$(x + 5)^2 - 9 = 0$$

Step 4: Shift the constant to the right side:

$$(x + 5)^2 = 9$$

Step 5: Take the square root of both sides and rearrange to show the solutions:

$$x + 5 = \pm 3$$

$$\therefore x = -2 \text{ or } x = -8$$

Notice that these are the same solutions that you would get by factorising:

$$x^2 + 10x + 16 = (x + 2)(x + 8) = 0 \Rightarrow x = -2 \text{ or } x = -8$$

ISBN: 9780170447058

2 Show that $2x^2 - 12x + 19$ is positive for all values of x.

Extra step: Factorise the first two terms:

$$2x^2 - 12x + 19 = 2(x^2 - 6x) + 19$$

Step 1: Using the coefficient of the x term, create an extra constant in order to produce a perfect square:

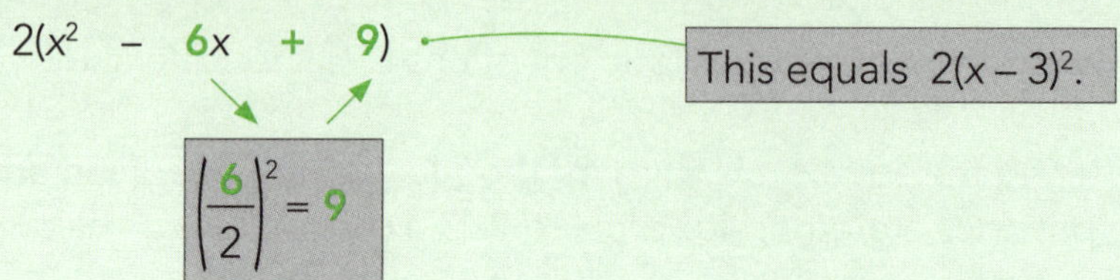

Step 2: Because you have added an extra **2 x 9 = 18**, you need to subtract 18:

$$2(x^2 - 6x + 9) - 18 + 19$$

Step 3: Rewrite the expression as a perfect square and a single constant:

$$2(x - 3)^2 - 18 + 19 = 2(x - 3)^2 + 1$$

So, $2x^2 - 12x + 19 = 2(x - 3)^2 + 1$

Squares are always positive, so $2(x - 3)^2 + 1$ must be positive.

$\therefore\ 2x^2 - 12x + 19$ must always be positive.

Consequently, the entire graph will be above the x-axis.

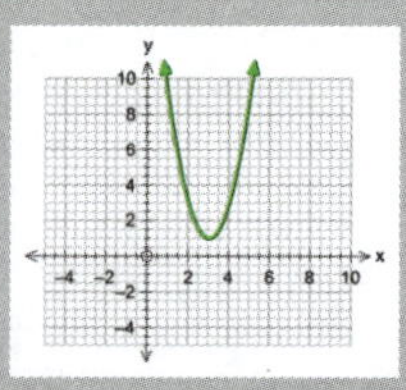

Find the value that completes the square, and then factorise each expression completely; include a perfect square in your answer.

1 $x^2 + 4x +$ ______________

____

____

2 $x^2 - 10x +$ ______________

____

____

3 $x^2 - 8x +$ ______________

____

____

4 $x^2 + 24x +$ ______________

____

____

5 $2(x^2 + 6x +$ ______________ $)$

____

____

6 $3(x^2 - 14x +$ ______________ $)$

____

____

ISBN: 9780170447058

Solve these equations by completing the square. Where appropriate you may leave your answer in the form $a \pm b\sqrt{c}$.

7 $x^2 + 2x - 35 = 0$

8 $x^2 - 2x - 3 = 0$

9 $x^2 - 7x - 18 = 0$

10 $x^2 - 12x + 23 = 0$

11 $x^2 - 8x + 6 = 0$

12 $x^2 + 3x + \frac{3}{4} = 0$

13 $4x^2 + 16x - 65 = 0$

14 $2x^2 - 3x - 35 = 0$

15 $3x^2 + 3x - 4\frac{1}{4} = 0$

16 $2x^2 - 3x - 1 = 0$

ISBN: 9780170447058

Write each expression in the form $(x \pm a)^2 + b$ in order to show that they are positive for all values of x.

17 $x^2 - 2x + 2$

18 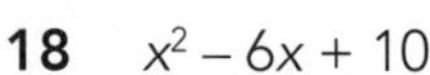 $x^2 - 6x + 10$

19  $x^2 - 10x + 30$

20 $x^2 - x + 5$

21 $2x^2 - 8x + 11$

22 $3x^2 - 24x + 49$

23 $5x^2 - 30x + 47$

24 $9x^2 - 72x + 145$

ISBN: 9780170447058

Long division

Polynomials can be factorised using a process that is similar to long division with numbers.

Method: Divide $6x^3 + 11x^2 - 3x - 2$ by the factor $3x + 1$.

Step 1:

$$\begin{array}{r|l} & 2x^2 \\ \hline 3x+1 & 6x^3 + 11x^2 - 3x - 2 \end{array}$$

Divide the first term of the factor (**3x**) into the first term of the expression to be factorised (**$6x^3$**). Write your answer (**$2x^2$**) on the top line.

Step 2:

$$\begin{array}{r|l} & 2x^2 \\ \hline 3x+1 & 6x^3 + 11x^2 - 3x - 2 \\ & 6x^3 + 2x^2 \end{array}$$

Multiply each term of the factor by your answer (**$2x^2$**), and write the result in the line below, matching the exponents.

Step 3:

$$\begin{array}{r|l} & 2x^2 \\ \hline 3x+1 & 6x^3 + 11x^2 - 3x - 2 \\ & 6x^3 + 2x^2 \\ \hline & \quad 9x^2 - 3x \end{array}$$

a Subtract **$6x^3 + 2x^2$** from **$6x^3 + 11x^2$** and write the result (**$9x^2$**) underneath.

b 'Bring down' the next term (**$-3x$**) and write it beside the **$9x^2$**.

Repeat step 1:

$$\begin{array}{r|l} & 2x^2 + 3x \\ \hline 3x+1 & 6x^3 + 11x^2 - 3x - 2 \\ & 6x^3 + 2x^2 \\ \hline & \quad 9x^2 - 3x \end{array}$$

Divide the first term of the factor (**3x**) into **$9x^2$**. Write your answer (**3x**) on the top line beside the $2x^2$.

Repeat step 2:

$$\begin{array}{r|l} & 2x^2 + 3x \\ \hline 3x+1 & 6x^3 + 11x^2 - 3x - 2 \\ & 6x^3 + 2x^2 \\ \hline & \quad 9x^2 - 3x \\ & \quad 9x^2 + 3x \end{array}$$

Multiply both terms of the factor by your answer (**3x**), and write the result in the line below, matching the exponents.

Repeat step 3:

$$\begin{array}{r|l} & 2x^2 + 3x \\ \hline 3x+1 & 6x^3 + 11x^2 - 3x - 2 \\ & 6x^3 + 2x^2 \\ \hline & \quad 9x^2 - 3x \\ & \quad 9x^2 + 3x \\ \hline & \qquad -6x - 2 \end{array}$$

a Subtract **$9x^2 + 3x$** from **$9x^2 - 3x$** and write the result (**$-6x$**) underneath.

b 'Bring down' the next term (**-2**) and write it beside the **$-6x$**.

Repeat step 1:

$$\begin{array}{r l} & 2x^2 + 3x - 2 \\ 3x + 1 \,\big) & 6x^3 + 11x^2 - 3x - 2 \\ & 6x^3 + 2x^2 \\ & \quad\; 9x^2 - 3x \\ & \quad\; 9x^2 + 3x \\ & \qquad\;\; -6x - 2 \end{array}$$

Divide the first term of the factor ($3x$) into $-6x$. Write your answer (-2) on the top line beside the $2x^2 + 3x$.

Repeat step 2:

$$\begin{array}{r l} & 2x^2 + 3x - 2 \\ 3x + 1 \,\big) & 6x^3 + 11x^2 - 3x - 2 \\ & 6x^3 + 2x^2 \\ & \quad\; 9x^2 - 3x \\ & \quad\; 9x^2 + 3x \\ & \qquad\;\; -6x - 2 \\ & \qquad\;\; -6x - 2 \end{array}$$

Multiply both terms of the factor by your answer (-2), and write the result in the line below, matching the exponents.

Repeat step 3:

$$\begin{array}{r l} & 2x^2 + 3x - 2 \\ 3x + 1 \,\big) & 6x^3 + 11x^2 - 3x - 2 \\ & 6x^3 + 2x^2 \\ & \quad\; 9x^2 - 3x \\ & \quad\; 9x^2 + 3x \\ & \qquad\;\; -6x - 2 \\ & \qquad\;\; -6x - 2 \\ & \qquad\qquad\;\; 0 \end{array}$$

Subtract $-6x - 2$ from $-6x - 2$ and write the result (0) underneath.

The 0 tells you that $3x + 1$ divides **exactly** into $6x^3 + 11x^2 - 3x - 2$ with **no remainder**.

So $\dfrac{6x^3 + 11x^2 - 3x - 2}{3x + 1} = 2x^2 + 3x - 2$ with no remainder.

Then in order to fully factorise $6x^3 + 11x^2 - 3x - 2$, use your Year 11 factorising skills to factorise $\mathbf{2x^2 + 3x - 2}$:

$$2x^2 + 3x - 2 = (2x - 1)(x + 2)$$

$$\therefore\; \mathbf{6x^3 + 11x^2 - 3x - 2 = (3x + 1)(2x - 1)(x + 2)}$$

ISBN: 9780170447058

Examples:

1 Divide $8x^3 + 26x^2 + 17x - 6$ by $2x + 3$, and use your result to help you complete the factorisation.

$$
\begin{array}{r|l}
 & 4x^2 + 7x - 2 \\
\hline
2x + 3 & 8x^3 + 26x^2 + 17x - 6 \\
 & \underline{8x^3 + 12x^2} \\
 & \quad\quad 14x^2 + 17x \\
 & \quad\quad \underline{14x^2 + 21x} \\
 & \quad\quad\quad\quad -4x - 6 \\
 & \quad\quad\quad\quad \underline{-4x - 6} \\
 & \quad\quad\quad\quad\quad\quad 0
\end{array}
$$

$\therefore 8x^3 + 26x^2 + 17x - 6 = (2x + 3)(4x^2 + 7x - 2)$
$= (2x + 3)(4x - 1)(x + 2)$

2 Divide $4x^3 - 7x^2 - 11x + 8$ by $4x + 5$.

$$
\begin{array}{r|l}
 & x^2 - 3x + 1 \\
\hline
\mathbf{4x + 5} & 4x^3 - 7x^2 - 11x + 8 \\
 & \underline{4x^3 + 5x^2} \\
 & \quad\quad -12x^2 - 11x \\
 & \quad\quad \underline{-12x^2 - 15x} \\
 & \quad\quad\quad\quad 4x + 8 \\
 & \quad\quad\quad\quad \underline{4x + 5} \\
 & \quad\quad\quad\quad\quad\quad \mathbf{3}
\end{array}
$$

Note the remainder of 3.

$$\therefore \frac{4x^3 - 7x^2 - 11x + 8}{4x + 5} = x^2 - 3x + 1 + \frac{\mathbf{3}}{\mathbf{4x + 5}}$$

3 Divide $9x^3 - x - 2$ by $3x - 2$.

Notice that there is **no x^2 term**. When this occurs, you **must** include the missing term in your division calculation, but write it as $\mathbf{0x^2}$.

$$
\begin{array}{r|l}
 & 3x^2 + 2x + 1 \\
\hline
3x - 2 & 9x^3 + \mathbf{0x^2} - x - 2 \\
 & \underline{9x^3 - 6x^2} \\
 & \quad\quad 6x^2 - x \\
 & \quad\quad \underline{6x^2 - 4x} \\
 & \quad\quad\quad\quad 3x - 2 \\
 & \quad\quad\quad\quad \underline{3x - 2} \\
 & \quad\quad\quad\quad\quad\quad 0
\end{array}
$$

$\therefore 9x^3 - x - 2 = (3x - 2)(3x^2 + 2x + 1)$

4 If $\dfrac{\mathbf{2x^4} + 7x^3 - 4x^2 - 27x - 20}{x - 2}$

$= 2x^3 + ax^2 + bx + c + \dfrac{d}{x - 2}$

find the values of a, b, c and d.

Notice that this polynomial includes a term with $\mathbf{x^4}$. Use the same method as you use with those with x^3.

$$
\begin{array}{r|l}
 & 2x^3 + 11x^2 + 18x + 9 \\
\hline
x - 2 & 2x^4 + 7x^3 - 4x^2 - 27x - 20 \\
 & \underline{2x^4 - 4x^3} \\
 & \quad\quad 11x^3 - 4x^2 \\
 & \quad\quad \underline{11x^3 - 22x^2} \\
 & \quad\quad\quad\quad 18x^2 - 27x \\
 & \quad\quad\quad\quad \underline{18x^2 - 36x} \\
 & \quad\quad\quad\quad\quad\quad 9x - 20 \\
 & \quad\quad\quad\quad\quad\quad \underline{9x - 18} \\
 & \quad\quad\quad\quad\quad\quad\quad\quad -2
\end{array}
$$

$\therefore a = 11, b = 18, c = 9$ and $d = -2$

ISBN: 9780170447058

Carry out the following long divisions.

1 Divide $x^3 + 8x^2 + 17x + 10$ by $x + 1$.

2 Divide $x^3 + 3x^2 + 5x - 9$ by $x - 1$.

3 Divide $4x^3 + 50x^2 + 105x + 50$ by $x + 10$.

4 Divide $x^3 - 10x^2 + 20x - 11$ by $x - 1$.

5 Divide $4x^3 - 13x + 6$ by $2x - 1$.

6 Divide $x^4 + x^3 - 7x^2 - x + 6$ by $x + 3$.

ISBN: 9780170447058

7 Divide $x^3 - 4x^2 + x + 10$ by $x - 2$.

8 Divide $2x^3 - 3x^2 - 19x + 25$ by $x - 3$.

Factorise the following completely, after using long division.

9 $6x^3 - 29x^2 + 21x - 4$, where one factor is $(2x - 1)$.

10 $x^3 - 28x - 48$, where one factor is $(x - 6)$.

11 $12x^3 + 4x^2 - 17x + 6$, where one factor is $(3x - 2)$.

12 $27x^3 - 54x^2 + 36x - 8$, where one factor is $(3x - 2)$.

ISBN: 9780170447058

Answer the following.

13 If $\frac{2x^3 - 3x^2 - 19x + 36}{x - 3} = ax^2 + bx + c + \frac{d}{x - 3}$

find the values of a, b, c and d.

14 If $\frac{3x^4 - 12x + 5}{x + 1} = 3x^3 + ax^2 + bx + c + \frac{d}{x + 1}$

find the values of a, b, c and d.

15 Use long division to simplify the expression $\frac{2x^3 - 5x^2 - 4x + 3}{x^2 - 2x - 3}$.

16 Use long division to simplify the expression $\frac{2x^4 + 7x^3 - 4x^2 - 27x - 18}{x^2 + x - 6}$.

 ISBN: 9780170447058

Factor and remainder theorems

1 The factor theorem

- You already know that in order to solve quadratic equations, you can factorise and then consider what happens when each factor is equated with 0.

Example: Solve $x^2 + 3x - 10 = 0$

$$(x - 2)(x + 5) = 0$$

$$\therefore \text{ either } (x - 2) = 0 \text{ or } (x + 5) = 0$$

So, we know that $x = 2$ and $x = -5$ both satisfy the equation $x^2 + 3x - 10 = 0$.

So, if $f(x) = x^2 + 3x - 10$, then $f(2) = 0$ and $f(-5) = 0$.

The factor theorem turns this around and says:

if $f(2) = 0$, then $(x - 2)$ is a factor of $x^2 + 3x - 10$

and if $f(-5) = 0$, then $(x + 5)$ is a factor of $x^2 + 3x - 10$.

In general terms:

For any polynomial:
If $f(a) = 0$, then $(x - a)$ is a factor of $f(x)$.
If $f(-a) = 0$, then $(x + a)$ is a factor of $f(x)$.

Note:
- Long division **can** be used to factorise polynomials, but you will not usually be given one of the factors to start with, so it is likely to take much longer.
- Use of the factor theorem, along with careful consideration of which values to substitute, is usually quicker.

Examples: Factorise the following using the factor theorem.

1 $x^3 + 4x^2 + x - 6$

> Carefully consider what numbers to substitute:
> 1 The numbers in the brackets must multiply to produce -6 ∴ start with factors of 6: 1, 2, 3 and 6.
> 2 Make the substitutions as easy for yourself as you can: begin with ±1, then ±2, etc.

$f(1) = (1)^3 + 4(1)^2 + (1) - 6 = 0 \Rightarrow$ $(x - 1)$ is a factor.

$f(-1) = (-1)^3 + 4(-1)^2 + (-1) - 6 = -1 + 4 - 1 - 6 \neq 0 \Rightarrow (x + 1)$ is not a factor.

$f(2) = (2)^3 + 4(2)^2 + (2) - 6 = 8 + 16 + 2 - 6 \neq 0 \Rightarrow (x - 2)$ is not a factor.

$f(-2) = (-2)^3 + 4(-2)^2 + (-2) - 6 = -8 + 16 - 2 - 6 = 0 \Rightarrow$ $(x + 2)$ is a factor.

Now, if $(x - 1)(x + 2)(x + a) = x^3 + 4x^2 + x - 6$, then $-2a = -6 \Rightarrow a = +3$.

So $\mathbf{x^3 + 4x^2 + x - 6 = (x - 1)(x + 2)(x + 3)}$.

ISBN: 9780170447058

2 $3x^3 - 10x^2 - 27x$ **+ 10**

Try ±1, ±2 and ±5.

$f(1) = 3(1)^3 - 10(1)^2 - 27(1) + 10 = 3 - 10 - 27 + 10 \neq 0 \Rightarrow (x - 1)$ is not a factor.

$f(-1) = 3(-1)^3 - 10(-1)^2 - 27(-1) + 10 = -3 - 10 + 27 + 10 \neq 0 \Rightarrow (x + 1)$ is not a factor.

$f(2) = 3(2)^3 - 10(2)^2 - 27(2) + 10 = 24 - 40 - 54 + 10 \neq 0 \Rightarrow (x - 2)$ is not a factor.

$f(-2) = 3(-2)^3 - 10(-2)^2 - 27(-2) + 10 = -24 - 40 + 54 + 10 = 0 \Rightarrow$ **$(x + 2)$ is a factor**.

$f(5) = 3(5)^3 - 10(5)^2 - 27(5) + 10 = 375 - 250 - 135 + 10 = 0 \Rightarrow$ **$(x - 5)$ is a factor**.

Now, if $(x$ **+ 2**$)$ and $(x$ **– 5**$)$ are factors of **3**$x^3 - 10x^2 - 27x$ **+ 10**, then the third factor must be **$(3x - 1)$**.

So $(x + 2)(x - 5) = x^2 - 3x - 10$ is a factor.

So $\mathbf{3x^3 - 10x^2 - 27x + 10 = (x + 2)(x - 5)(3x - 1)}$.

Use the factor theorem to help you factorise the following.

1 $x^3 - 2x^2 - x + 2$

2 $x^3 + x^2 - 4x - 4$

 ISBN: 9780170447058

3 $x^3 + x^2 - 8x - 12$

4 $2x^3 + x^2 - 13x + 6$

5 $3x^3 + 17x^2 + 18x - 8$

ISBN: 9780170447058

6 $4x^3 + 3x^2 - 25x + 6$

7 $2x^3 + 11x^2 + 2x - 15$

8 $6x^3 + 65x^2 + 133x - 24$

ISBN: 9780170447058

2 The remainder theorem

- This provides a quick method of finding the remainder when a polynomial is divided by a factor.

In general terms:

For any polynomial:
If it is divided by (x + a), then the remainder is f(–a).
If it is divided by (x – a), then the remainder is f(a).

Note:

- Long division **can** be used to find the remainder, but use of the remainder theorem will be much faster.

Examples:

1 Find the remainder when $x^3 + 4x^2 + x - 6$ is divided by $(x + 1)$.

$f(-1) = (-1)^3 + 4(-1)^2 + (-1) - 6 = -1 + 4 - 1 - 6$

$= -4 \Rightarrow \frac{x^3 + 4x^2 + x - 6}{x + 1}$ has a remainder of -4.

2 Find the remainder when $2x^3 + x^2 - 13x + 6$ is divided by $(x - 5)$.

$f(5) = 2(5)^3 + (5)^2 - 13(5) + 6 = 250 + 25 - 65 + 6$

$= 216 \Rightarrow \frac{2x^3 + x^2 - 13x + 6}{x - 5}$ has a remainder of 216.

3 Dividing $x^4 - 3x^3 + ax^2 - 8$ by $(x + 2)$ gives a remainder of 48. What is the value of a?

$$f(-2) = (-2)^4 - 3(-2)^3 + a(-2)^2 - 8 = 48$$

$$16 + 24 + 4a - 8 = 48$$

$$a = 4$$

Answer the following.

9 What is the remainder when $x^3 - 2x^2 - x + 2$ is divided by $(x - 3)$?

---

10 What is the remainder when $x^3 + x^2 - 4x - 4$ is divided by $(x + 4)$?

ISBN: 9780170447058

11 What is the remainder when $2x^3 + x^2 - 13x + 6$ is divided by $(x + 2)$?

12 What is the remainder when $3x^3 + 17x^2 + 18x - 8$ is divided by $(x + 5)$?

13 Dividing $x^3 + ax^2 - 8x - 12$ by $(x - 4)$ gives a remainder of 36. What is the value of a?

14 Dividing $4x^3 + 3x^2 + ax + 6$ by $(x + 1)$ gives a remainder of 30. What is the value of a?

15 Given that $(x + 5)$ is a factor of $2x^3 + px^2 + 2x - 15$, find the value of p.

ISBN: 9780170447058

16 Find the value of p if $(x - 7)$ is a factor of $3x^3 + px^2 - 9x + 14$.

17 One solution of the equation $5x^3 + 9x^2 + px + 8 = 0$ is $x = -4$. Find the value of p and the other two solutions to the equation.

18 If $(x + 3)$ and $(x - 2)$ are both factors of $2x^3 + px^2 + qx + 30$, find the values of p and q.

19 If $(x - 4)$ and $(x + 2)$ are both factors of $4x^3 + px^2 + qx + 24$, find the values of p and q.

ISBN: 9780170447058

Complex numbers

Imaginary numbers

Consider the function $y = x^2 - 4$: to find 'solutions' (i.e. points where it crosses the x-axis), we let $y = 0$, and solve $x^2 = 4 \Rightarrow x = \sqrt{4} = \pm 2$.

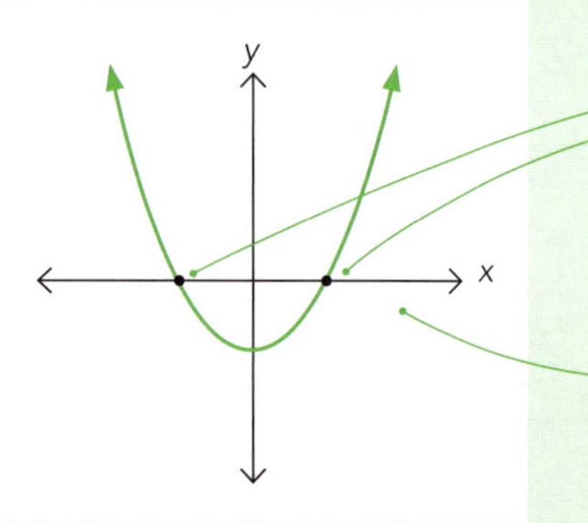

'Solutions' occur at $x = \pm 2$.

These solutions are on the x-axis, which includes all the **real numbers**.

Consider the function $y = x^2 + 4$: there are no 'solutions' (i.e. points where it crosses the x-axis). If we let $y = 0$, and solve $x^2 = -4 \Rightarrow$ **$x = \sqrt{-4}$**.

So far you have not met any method for dealing with this equation.

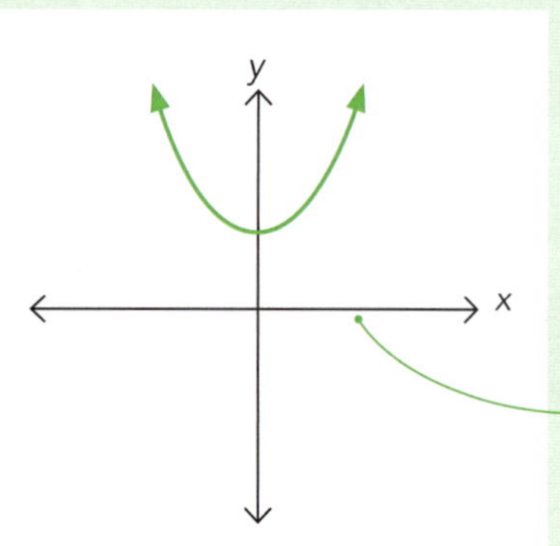

No solutions along the x-axis, which contains all the real numbers.

In order to deal with $x = \sqrt{-4}$, we consider the imaginary number $i = \sqrt{-1}$:

$$\begin{aligned} x &= \sqrt{4 \times -1} \\ &= \sqrt{2^2 \times -1} \\ &= \sqrt{2^2} \times \sqrt{-1} \\ &= 2i \end{aligned}$$

This means that $i^2 = -1$.

In general: $$\begin{aligned} \sqrt{-k^2} &= \sqrt{k^2 \times -1} \\ &= \sqrt{k^2} \times \sqrt{-1} \\ &= ki, \text{ where } i = \sqrt{-1} \end{aligned}$$

 ISBN: 9780170447058

Manipulation of imaginary numbers

Addition and subtraction

- These can be added and subtracted by considering the imaginary numbers as 'like' terms.

Examples: **1** $9i - 2i = 7i$

2 $i - \sqrt{3}\,i = i(1 - \sqrt{3})$

Multiplication and division

- When two imaginary numbers are **multiplied** or **divided**, the result is **a real number**.

Examples: **1** $3i \times 5i = 15i^2$ — But $i^2 = -1$.

$= -15$

2 $12i \div 3i = 4$

Powers of imaginary numbers

Examples:

1 $i^3 = (i^2) \times i$
$= (-1) \times i$
$= -i$

2 $i^4 = (i^2)^2$
$= (-1)^2$
$= 1$ — Useful to remember that $i^{4n} = 1$.

3 $i^5 = i^4 \times i$
$= 1 \times i$
$= i$

4 $i^{-1} = \frac{i}{i^2}$
$= \frac{i}{-1}$
$= -i$

Calculate the following.

1 $3i + 7i =$ ________________

2 $\sqrt{8}\,i - 5i =$ ________________

3 $2i \times 13i =$ ________________
____

4 $24i \div 8i =$ ________________
____

5 $i^7 =$ ________________
____

6 $i^9 =$ ________________
____

7 $(2i)^7 =$ ________________
____

8 $i^{-3} =$ ________________
____

Note: You can use your graphics calculator for these calculations: **Shift 0 = *i***

Complex numbers — the basics

Complex number are made up of a **real part** and an **imaginary part**.
Complex numbers are usually written as:

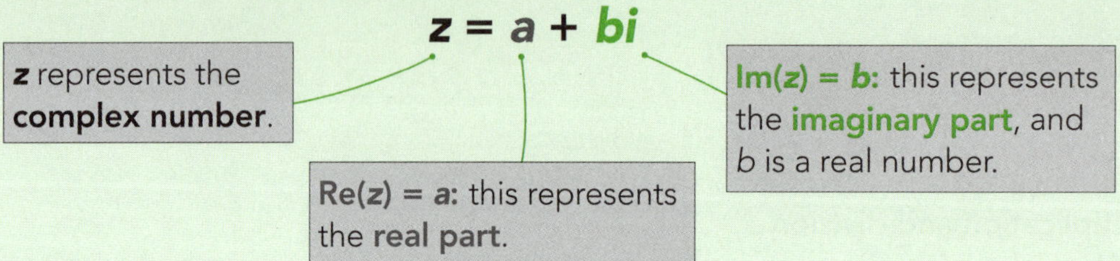

Examples: Write down the real and imaginary parts of the following complex numbers.

1 $z = 4 + 3i$
Re(z) = 4
Im(z) = 3

2 $z = 9i$
Re(z) = 0
Im(z) = 9

3 $z = 7$
Re(z) = 7
Im(z) = 0

4 $z = 2 - xi + y + 5i$
Re(z) = $2 + y$
Im(z) = $5 - x$

Equate the real and imaginary parts of the following complex numbers in order to find the values of x and y.

5 $x - 5i = 4 + yi$
Re(z): $x = 4$
Im(z): $y = -5$

6 $x + 3iy - 2 + i = 3 + 7i$
Re(z): $x - 2 = 3$ $\Rightarrow x = 5$
Im(z): $3y + 1 = 7$ $\Rightarrow y = 2$

7 $2x + 12i = 8 - 3(y + 2)i$
Re(z): $2x = 8$ $\Rightarrow x = 4$
Im(z): $12 = -3(y + 2)$ $\Rightarrow y = -6$

8 $-8i(x + yi) = 24i$
$x + yi = -3$
Re(z): $x = -3$
Im(z): $y = 0$

Write down the real and imaginary parts of the following complex numbers.

1 $z = 2 + 9i$
Re(z) = ____________________
Im(z) = ____________________

2 $z = 27$
Re(z) = ____________________
Im(z) = ____________________

3 $z = 15i$
Re(z) = ____________________
Im(z) = ____________________

4 $z = x - yi$
Re(z) = ____________________
Im(z) = ____________________

 ISBN: 9780170447058

5 $z = 4 - (x + 2)i$

Re(z) = ______

Im(z) = ______

6 $z = 6 + i - 2y + xi$

Re(z) = ______

Im(z) = ______

7 $z = 13i - 9(xi - 1)$

Re(z) = ______

Im(z) = ______

8 $z = 3(2i + x) - (5 + 2y)i$

Re(z) = ______

Im(z) = ______

Equate the real and imaginary parts of the following complex numbers in order to find the values of x and y.

9 $x - 2i = -5 + yi$

10 $xi + 15 = 3y - i$

11 $x + 12i = (y - 1)i$

12 $2x - yi + 3 = 6i - 1$

13 $5x + i(y + 2) = 3yi$

14 $i(3x + i) - 7 = 9i + 4y$

15 $i(3 + 5xi) = i(y + 2) + 25$

16 $-2i(ix + 3) = 5x - 2(3 - yi)$

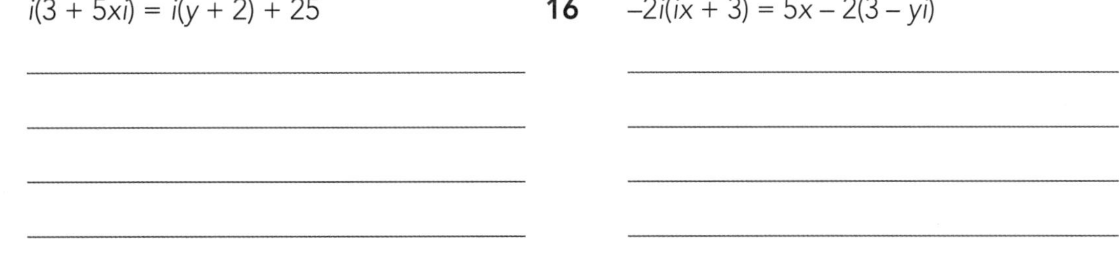

Complex numbers and Argand diagrams

- Complex numbers can be plotted on a graph called an **Argand diagram**.

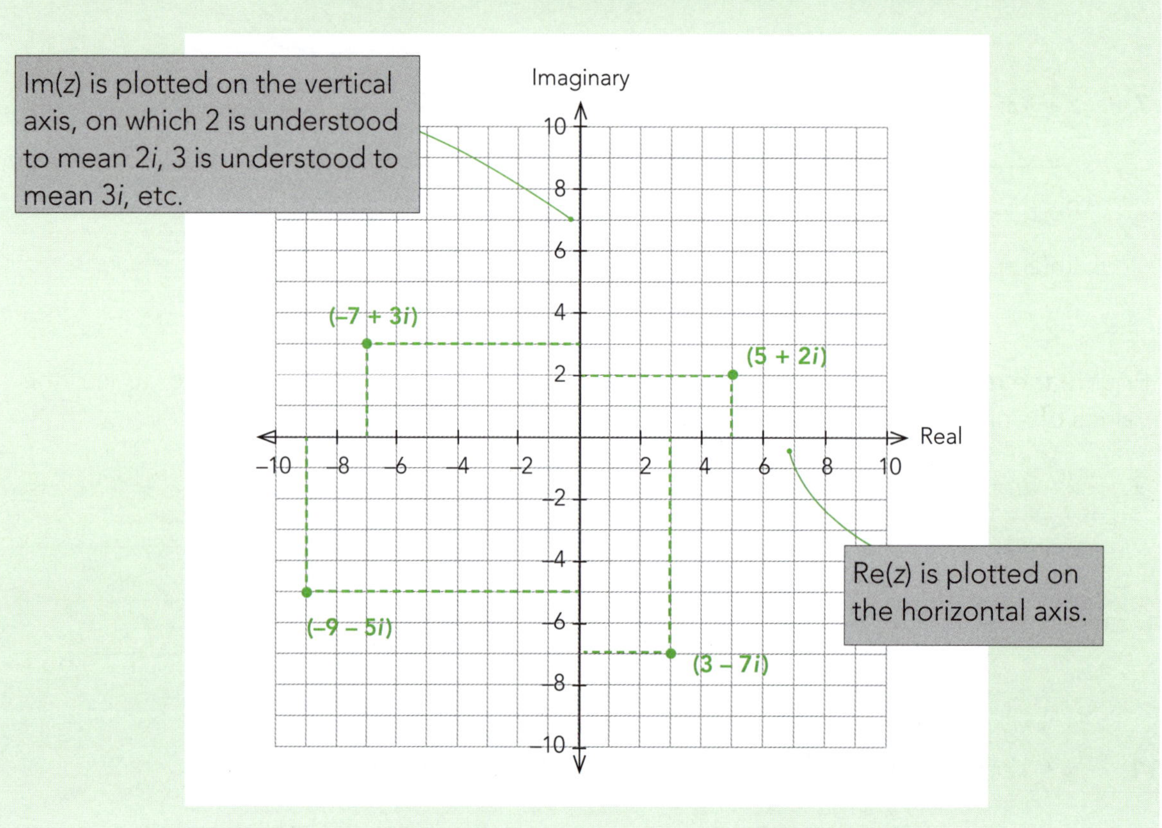

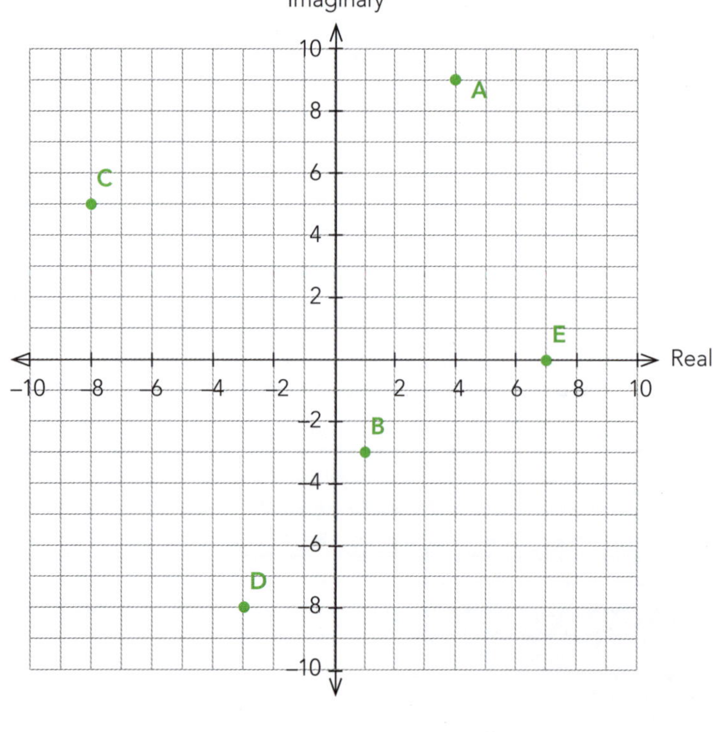

Write the complex numbers represented by the labelled points.

A ________________

B ________________

C ________________

D ________________

E ________________

Draw and label the following complex numbers.

$F = 4 + i$

$G = -2 + 9i$

$H = 8 - 7i$

$I = -6 - 5i$

$J = -9i$

 ISBN: 9780170447058

Manipulation of complex numbers

Addition and subtraction

- Add or subtract the real parts of the complex number (consider these as 'like' terms).
- Then add or subtract the imaginary parts (consider these as 'like' terms).
- These can also be drawn on an Argand diagram like vectors.

Examples:

1 $(-7 + 3i) + (4 + 6i) = (-7 + 4) + (3i + 6i)$
$= -3 + 9i$

Notice that order doesn't matter:

$(4 + 6i) + (-7 + 3i) = (4 + -7) + (6i + 3i)$
$= -3 + 9i$

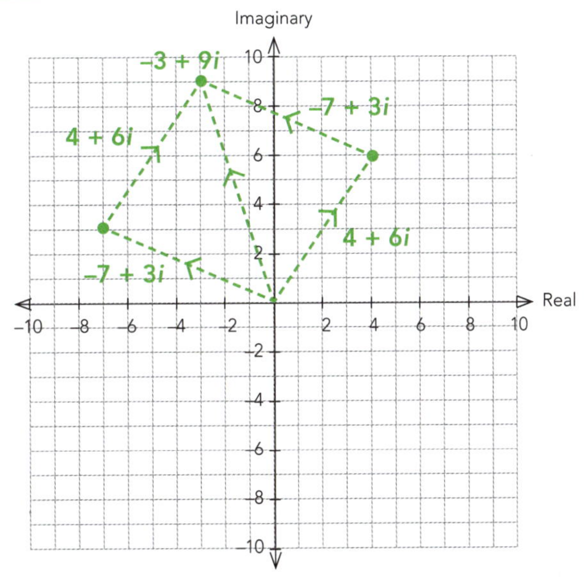

2 $(1 - 5i) - (6 - 2i) = (1 - 5i) + (-6 + 2i)$
$= (1 - 6) + (-5i + 2i)$
$= -5 - 3i$

Notice that $6 - 2i$ is the same length as, but in the opposite direction of, $-6 + 2i$.

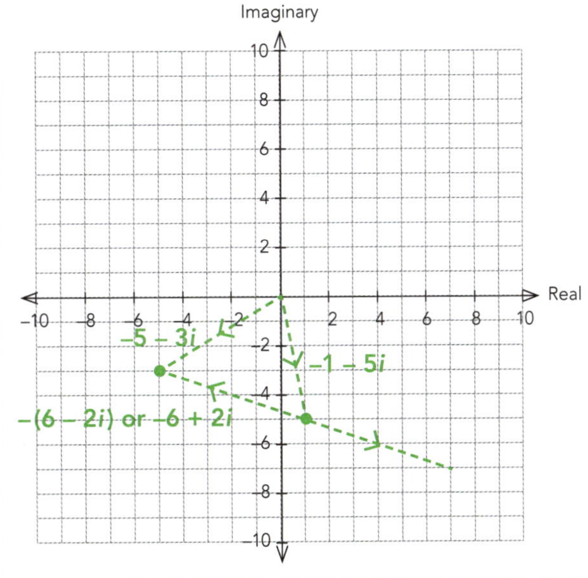

ISBN: 9780170447058

Answer the following.

1 $(5 + 6i) + (2 + 3i) =$

2 $(1 - 4i) + (2 - i) =$

3 $(-5) + (2 - i) =$

4 $(-2 - 3i) + (-6i) =$

5 $(9 + 2i) - (2 + 4i) =$

6 $(-7 + i) - (-5 + 3i) =$

7 $(8i) - (1 - 3i) =$

8 $(-1 - 6i) - (-2 - i) =$

9 $(x + yi) + (y - xi) =$

10 $(3x - 2iy) + (-x + i) =$

11 $2(x - yi) - (-x - yi) =$

12 $-(-x - 3yi) - (-2y - xi) =$

13 Complex numbers s and t are represented on the Argand diagram.

Show the following on the same diagram.

a $u = s + t$

b $v = s - t$

c $w = t - s$

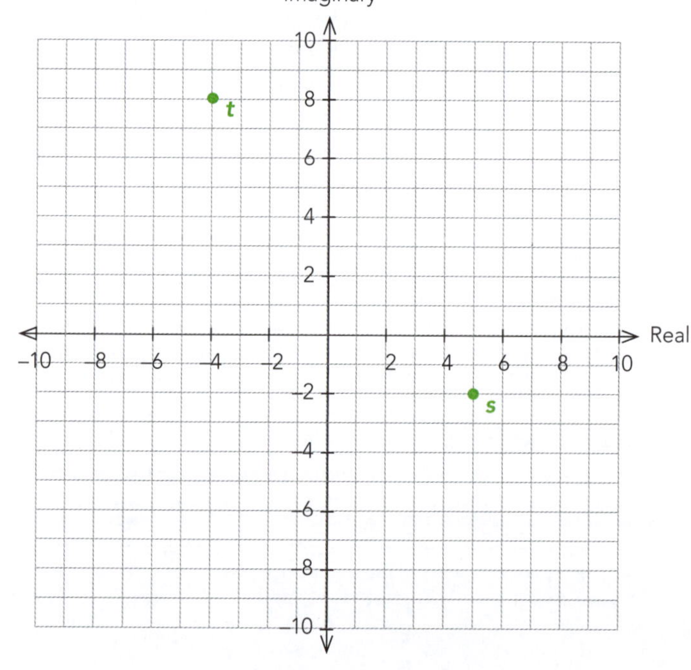

ISBN: 9780170447058

Multiplication

- Remember that $i^2 = -1$.
- Use the same method as for multiplying brackets.

Examples:

1 $(3-2i)(4+5i) = 12 + 15i - 8i - 10i^2$
$= 12 + 7i + 10$
$= 22 + 7i$

Always write your answer in **$a + bi$** form.

2 $(2-3i)^3 = (2-3i) \times (2-3i) \times (2-3i)$
$= (2-3i)(4 - 6i - 6i + 9i^2)$
$= (2-3i)(4 - 12i - 9)$
$= (2-3i)(-5 - 12i)$
$= -10 - 24i + 15i + 36i^2$
$= -10 - 24i + 15i - 36$
$= -46 - 9i$

Hint: You can use your graphics calculator for these calculations: **Shift 0 = *i***

Answer the following, and write your answers in $a + bi$ form.

1 $5(2-3i) =$ ____________

2 $3i(2+i) =$ ____________

3 $i(5-4i) =$ ____________

4 $-3i(7-2i) =$ ____________

5 $(3+2i)(1+4i) =$ ____________

6 $(2+3i)(1-6i) =$ ____________

7 $(1-8i)(2+3i) =$ ____________

8 $(5-6i)(2-i) =$ ____________

9 $(-2-3i)(7-5i) =$ ____________

10 $(-4-i)(-3-6i) =$ ____________

11 $(2+3i)^2 =$ ____________

12 $(3-5i)^2 =$ ____________

13 $(x+yi)(y+xi) =$ ____________

14 $(3x-2i)(4+xi) =$ ____________

ISBN: 9780170447058

15 $(y - 2iy)(5x - 3xi) =$ ________________

16 $(3x - 2iy)(-x + i) =$ ________________

17 $(\sqrt{5} - i)(\sqrt{5} + i) =$ ________________

18 $(2 + i)^3 =$ ________________

19 $(1 + \sqrt{3}i)^3 =$ ________________

20 $(7 + \sqrt{2}i)(7 - \sqrt{2}i) =$ ________________

21 $s = 3 - bi$ and $t = 4 + i$.
Find the value of b if $st = 14 - 5i$.

22 $s = 4 + \sqrt{3}i$ and $t = 4 - bi$.
Find the value of b if $st = 19$.

23 Solve the equation $(x + yi)(1 + i) = 2 + 8i$.

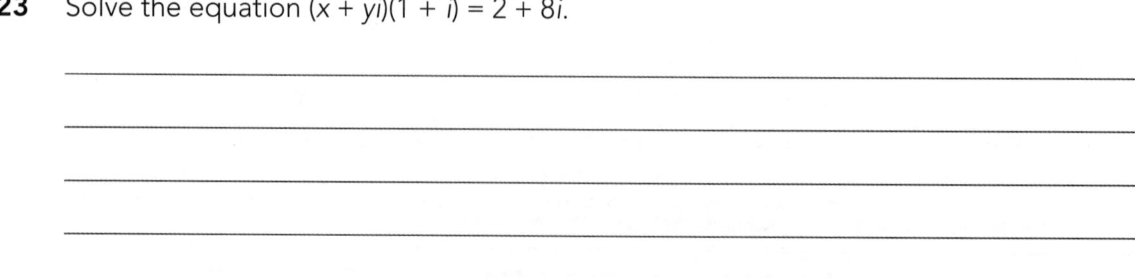

 ISBN: 9780170447058

Division

- Dividing one complex number by another results in another complex number.
- An answer with i in the denominator (e.g. $\frac{2}{3+i}$) is **not** acceptable.
- Use a trick that is similar to the one used in rationalising surds.
  Trick: Remember that $(a + b)(a - b) = a^2 - b^2$.

Notice that both of these terms are squares, so we can use this to get rid of a complex number (i) on the bottom of a fraction.

- **$(a + bi)$** and **$(a - bi)$** are known as the **conjugate complex numbers**.

$$(a + bi)(a - bi) = a^2 - (bi)^2$$
$$= a^2 - b^2(-1)$$
$$= a^2 + b^2$$

No complex number in this expression.

- If $z = a + bi$, then $a - bi = \bar{z}$.
- If $w = a - bi$, then $a + bi = \overline{w}$.

For conjugates, use the same letter, but add a line over it, e.g. z and $\bar{z}$ are conjugate complex numbers.

- Conjugate complex numbers on a graph are **reflections** of each other in the **x-axis**.

Examples:

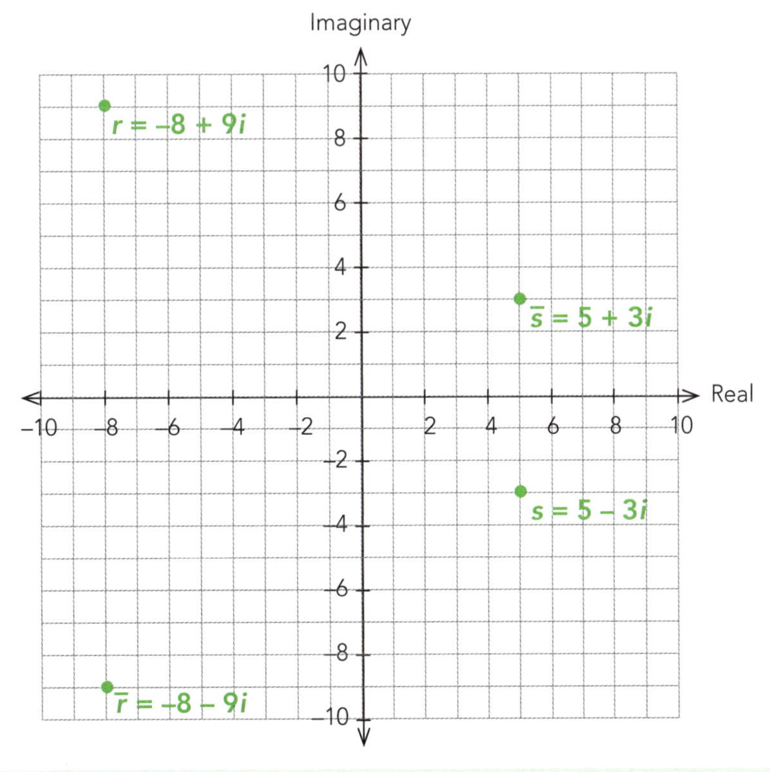

ISBN: 9780170447058

Examples: Rewrite the following expressions in the form $a \pm bi$.

1 $\dfrac{2-3i}{1+4i}$

$$\frac{2-3i}{1+4i} \times \frac{1-4i}{1-4i} = \frac{2-8i-3i+12i^2}{1-16i^2}$$
$$= \frac{2-11i+12(-1)}{1-16(-1)}$$
$$= \frac{-10-11i}{17}$$

$1^2 + 4^2 = 17$

$$= \frac{-10}{17} - \frac{11}{17}i$$

2 $\dfrac{5+2i}{-3+i}$

$$\frac{5+2i}{-3+i} \times \frac{-3-i}{-3-i} = \frac{-15-5i-6i-2i^2}{9-i^2}$$
$$= \frac{-15-2(-1)-11i}{9-(-1)}$$
$$= \frac{-13-11i}{10}$$

$(-3)^2 + 1^2 = 10$

$$= -1.3 - 1.1i$$

Answer the following.

1 If $p = 2 + 7i$, then $\bar{p} =$ ______________

2 If $s = 6 - 11i$, then $\bar{s} =$ ______________

3 If $q = 1 + 2i$, then $q \times \bar{q} =$

4 If $r = 5 - 3i$, then $r \times \bar{r} =$

5 If $u = a - 6i$, and $u \times \bar{u} = 40$, calculate the value of a.

6 If $v = 8 - bi$, and $v \times \bar{v} = 73$, calculate the value of b.

Rewrite the following in the form $a \pm bi$.

7 $\dfrac{1}{2+3i}$

8 $\dfrac{4}{5-2i}$

 ISBN: 9780170447058

9 $\dfrac{i}{3+2i}$

10 $\dfrac{2i}{1-5i}$

11 $\dfrac{1+3i}{5-2i}$

12 $\dfrac{4-3i}{2+i}$

13 $\dfrac{7-i}{3i-2}$

14 $\dfrac{2+3i}{7-i}$

Answer the following.

15 The complex number $\dfrac{2}{1-i}$ can be expressed in the form $k(1+i)$, where k is a real number. Find the value of k.

16 The complex number $\dfrac{-4-2i}{3-i}$ can be expressed in the form $k(1+i)$, where k is a real number. Find the value of k.

ISBN: 9780170447058

17 If $\frac{a+bi}{c+di} = \frac{p}{c^2+d^2} + \frac{q}{c^2+d^2}i$, write expressions for p and q in terms of a, b, c and d.

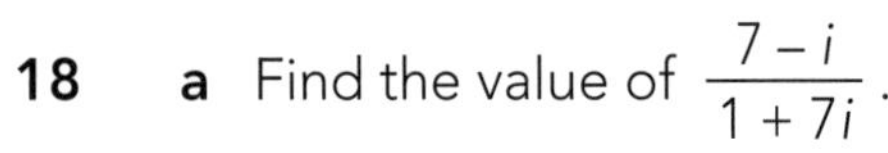

18 **a** Find the value of $\frac{7-i}{1+7i}$.

b Show that the expression $\frac{k-i}{1+ki} = i$ for any value of k.

19 Find all possible values of k that make $u = \frac{k+9i}{1+ki}$ a purely real number.

 ISBN: 9780170447058

Mixing it up

Answer the following.

1 **a** If $u = 3 + 2i$ and $v = 4i - 1$, find $\bar{u} - 2v$, giving your solution in the form $a + bi$.

b Find real numbers p and q such that $pu + qv = 12 + 22i$.

2 **a** If $w = 5 - 2i$ and $y = 4 + 3i$, find $\bar{w}^2 - wy$, giving your solution in the form $a + bi$.

b Find real numbers p and q such that $pw + qy = 47 - 5i$.

3 Express $\frac{i}{1 + 2i} - \frac{1}{2i}$ in the form $a + bi$, where a and b are real numbers.

ISBN: 9780170447058

4 Given that $w = 2 + i$, find the value of $\bar{w}^2 - \frac{w}{\bar{w}}$, giving your answer in the form $a + bi$, where a and b are real.

5 Given that $v = 1 + 3i$, find the value of $v^2 + \frac{1}{\bar{v}}$, giving your answer in the form $a + bi$, where a and b are real.

6 Given that $w = \sqrt{3} + 2i$, find the value of $\frac{w^2}{w\bar{w} - 2}$, giving your answer in the form $a + bi$, where a and b are real.

7 Given that $z = 2 - 3i$, find the value of $\bar{z}^2 - \frac{1}{z^2}$, giving your answer in the form $a + bi$, where a and b are real.

ISBN: 9780170447058

8 Write the complex number $10\left(\frac{4-i^6}{1-3i}\right)$ in the form $a + bi$, where a and b are real numbers.

9 Write the complex number $\left(\frac{2i+3i^9}{1-2i}\right)^2$ in the form $a + bi$, where a and b are real numbers.

10 If $(2-i)^4 = a + bi$, find the values of a and b.

11 If u and v are complex numbers, prove that $\overline{uv} = \overline{u} \times \overline{v}$.

ISBN: 9780170447058

The modulus

- The **modulus** of a complex number z is written **$|z|$ or mod(z)**.
- The **modulus** equals the **length** of the vector connecting the origin with the point (x, yi).
- By Pythagoras, if $z = x + yi$, then **$\text{mod}(z) = |z| = \sqrt{x^2 + y^2}$**.

This is on your formula sheet.

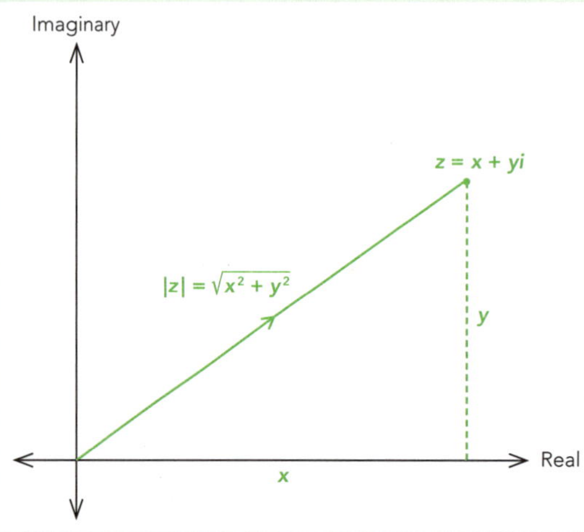

Examples:

1 Calculate the modulus of $2 - 2i$.

$$|2 - 2i| = \sqrt{2^2 + 2^2}$$
$$= \sqrt{8}$$
$$= 2\sqrt{2}$$

It is sometimes possible to simplify the surd.

2 If $v = 7 - 4i$ and $w = 2 + i$, calculate $|\overline{v} + \overline{w}|$.

$$|\overline{v} + \overline{w}| = |(7 + 4i) + (2 - i)|$$
$$= |9 + 3i|$$
$$= \sqrt{9^2 + 3^2}$$
$$= \sqrt{90}$$
$$= 3\sqrt{10}$$

Calculate the modulus of each of the following.

1 $2 + 3i$

2 6

3 $-5i$

4 $-1 - \sqrt{3}i$

5 $8 - 3i$

6 $2 + x + 3i$

ISBN: 9780170447058

For numbers **7–14**, let $u = 3 - 2i$, $v = -5 + i$ and $w = 1 + i$.

7 Calculate the value of $|u|$.

8 Calculate the value of $|\overline{v}|$.

9 Calculate the value of $|v + w|$.

10 Calculate the value of $|\overline{u - v}|$.

11 Calculate the value of $\frac{1}{|u|}$.

12 Calculate the value of $\frac{1}{|\overline{w}|}$.

13 Find the modulus of $\frac{u}{w}$.

14 Find mod $(\frac{1}{w^2})$.

15 If $p = \frac{2i}{1 + i} + 6$, find mod(p).

16 If $q = \frac{20}{2 + i} - 2i$, find mod($q$).

ISBN: 9780170447058

17 **a** If $z = a + bi$, prove that $|z| = |\bar{z}|$.

b Explain geometrically why $|z| = |\bar{z}|$.

18 If $z = a + bi$, prove that $|z|^2 = z\,\bar{z}$.

A challenge ...

19 If $z = a + bi$ and $w = c + di$, prove that $|zw| = |z||w|$.

 ISBN: 9780170447058

Polynomials again

The fundamental theorem of algebra

For any polynomial ax^n ± …… = 0, where n is a positive integer and the highest power of x, there are exactly n roots (solutions).

Examples: Quadratics (ax^2 ± …… = 0) have **2** roots.
Cubics (ax^3 ± …… = 0) have **3** roots.
Quartics (ax^4 ± …… = 0) have **4** roots, etc.

Note: Some roots may be repeated. For example: $x^2 - 10x + 25 = 0$ has **two equal** roots at $x = 5$.

Quadratics

- Remember that for quadratic equations that look like 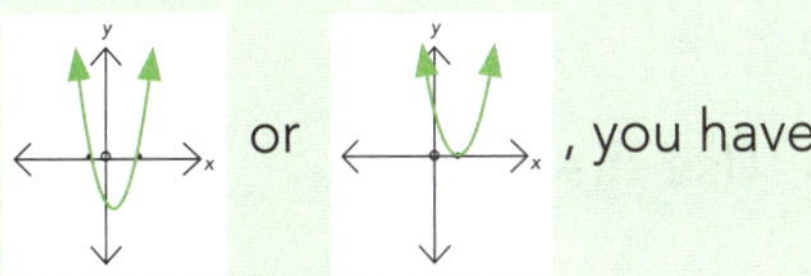 , you have a range of methods of finding the roots.
- Complex numbers enable us to find roots to quadratic equations that look like 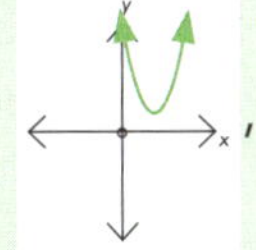 , because now you can deal with $\sqrt{\text{negative number}}$.
- Complex roots of any polynomial always come in **conjugate pairs**, e.g. $2 + i$ and $2 - i$.
- Three methods can be used:
  1. If $ax^2 + bx + c = 0$, and b = 0, then rearrange the equation so x^2 = number.
  2. Complete the square.
  3. Use the formula.

1 If b = 0, then rearrange the equation.

Example: 1 Solve the equation $x^2 + 4 = 0$.

$$x^2 = -4$$
$$x = \pm\sqrt{-4}$$
$$= \pm\sqrt{4i^2}$$
$$= \pm 2i$$

Remember that $-1 = i^2$.

Notice that, as with other quadratic equations, there are **two** solutions: $+2i$ and $-2i$. These are called **conjugate roots**.

Check the answer: $(x - 2i)(x + 2i) = x^2 - 4i^2$
$= x^2 + 4$ ✓

ISBN: 9780170447058

2 Complete the square (revise page 35 if you need to)

Example: Solve the equation $x^2 + 10x + 26 = 0$.

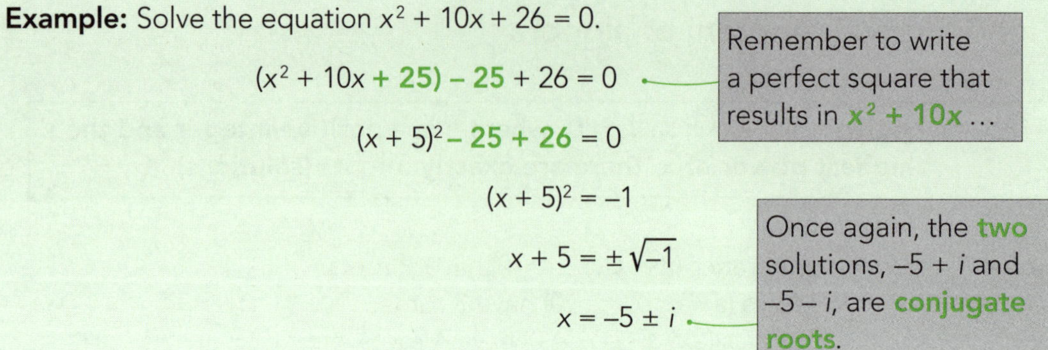

$$(x^2 + 10x + 25) - 25 + 26 = 0$$

$$(x + 5)^2 - 25 + 26 = 0$$

$$(x + 5)^2 = -1$$

$$x + 5 = \pm\sqrt{-1}$$

$$x = -5 \pm i$$

Remember to write a perfect square that results in $x^2 + 10x$ …

Once again, the **two** solutions, $-5 + i$ and $-5 - i$, are **conjugate roots**.

Check the answer: $(x - (-5 + i))(x - (-5 - i)) = (x + 5 - i)(x + 5 + i)$

$$= x^2 + 5x + xi + 5x + 25 + 5i - xi - 5i - i^2$$

$$= x^2 + 10x + 26 \quad ✓$$

3 Use the formula

Example: Solve the equation $x^2 - 6x + 12 = 0$.

$$x = \frac{-b \pm \sqrt{b^2 - 4ac}}{2a}$$

$$= \frac{-(-6) \pm \sqrt{(-6)^2 - 4 \times 1 \times 12}}{2a}$$

$$= \frac{6 \pm \sqrt{36 - 48}}{2}$$

$$= \frac{6 \pm \sqrt{-12}}{2}$$

$$= \frac{6 \pm \sqrt{12i^2}}{2}$$

$$= 3 \pm \frac{2\sqrt{3i^2}}{2}$$

$$= 3 \pm \sqrt{3}i$$

Remember that $-1 = i^2$.

Once again, the **two** solutions, $3 + \sqrt{3}i$ and $3 - \sqrt{3}i$, are **conjugate roots**.

Check the answer: $(x - (3 + \sqrt{3}i))(x - (3 - \sqrt{3}i)) = (x - 3 - \sqrt{3}i)(x - 3 + \sqrt{3}i)$

$$= x^2 - 3x + \sqrt{3}xi - 3x + 9 - 3(\sqrt{3}i) - \sqrt{3}xi + 3(\sqrt{3}i) - 3i^2$$

$$= x^2 - 6x + 12 \quad ✓$$

ISBN: 9780170447058

Solve the following.

1 $x^2 + 100 = 0$

2 $4x^2 + 100 = 0$

Solve the following either by completing the square or by using the quadratic formula.

3 $x^2 - 2x + 5 = 0$

4 $x^2 - 4x + 5 = 0$

5 $x^2 + 4x + 8 = 0$

6 $x^2 - 2x + 4 = 0$

7 $x^2 - 6x + 12 = 0$

8 $x^2 + x + 1 = 0$

ISBN: 9780170447058

9 Write $x^2 - 6x + 34$ as a product of two factors.

10 Write $3x^2 + 2x + 1$ as a product of two factors.

11 If $z = 2 + 3i$ is one solution to an equation $az^2 + bz + c = 0$, write down the other solution and find the values of a, b and c.

12 If $z = 3 + \sqrt{3}i$ is one solution to an equation $az^2 + bz + c = 0$, write down the other solution and find the values of a, b and c.

13 If $z = -1 + \sqrt{2}i$ is one solution to an equation $az^2 + bz + c = 0$, write down the other solution and find the values of a, b and c.

 ISBN: 9780170447058

Cubics

- Cubic equations can look like 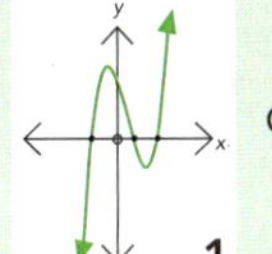 or 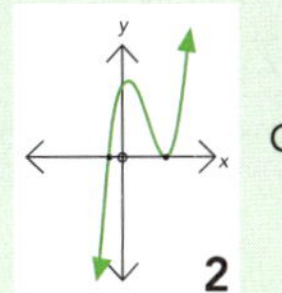 or 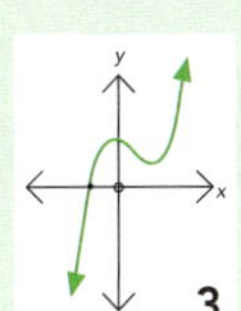 (or mirror images).

- Notice that cubic equations always have **at least one real root** because they must cross the x-axis.

- The third type of equation (**3**) has **one real root** and **a pair of conjugate roots**.

  Example: $x^3 - 7x^2 + 17x - 15 = 0$ has roots 3, $2 + i$ and $2 - i$.

As a result, $x^3 - 7x^2 + 17x - 15$ has three factors: $(x - 3)$, $(x - (2 + i))$ and $(x - (2 - i))$
or $(x - 3)$, $(x - 2 - i)$ and $(x - 2 + i)$.

- You have used long division and the factor theorem to solve the first two types of cubic equations (**1** and **2**).

- We can use the same methods, along with knowledge of complex numbers, to find solutions to the third type (**3**).

Examples:

1 One solution to the equation $z^3 + Az^2 + 6z - 20$ is $z = -1 + 3i$.
If A is a real number, find the value of A and the other two solutions to the equation.

One solution is $z = -1 + 3i \Rightarrow$ another solution is $z = -1 - 3i$.

$$(z - (-1 + 3i))(z - (-1 - 3i)) = (z + 1 - 3i)(z + 1 + 3i)$$

Multiply the two factors that you know.

$$= z^2 + z + 3zi + z + 1 + 3i - 3zi - 3i - 9i^2$$

$$= z^2 + 2z + 10$$

$z^2 + 2z + 10$ must be a factor of $z^3 + Az^2 + 6z - 20$.

Find the other solution (a):

$$(z^2 + 2z + 10)(z - a) = z^3 + Az^2 + 6z - 20$$

$+10 \times -a = -20 \Rightarrow a = 2$,
so $(z - 2)$ is a factor.

Find the value of A:

$$(z^2 + 2z + 10)(z - 2) = z^3 - 2z^2 + 2z^2 - 4z + 10z - 20$$

$$= z^3 + 0z^2 + 6z - 20$$

$\therefore$ A = 0 and the other two solutions are 2 and $-1 - 3i$.

ISBN: 9780170447058

2 One solution to the equation $2z^3 + Az^2 + 32z - 13$ is $z = 3 + 2i$.
If A is a real number, find the value of A and the other two solutions to the equation.

One solution is $z = 3 + 2i \Rightarrow$ another solution is $z = 3 - 2i$.

Multiply the two factors that you know.

$$(z - (3 + 2i))(z - (3 - 2i)) = (z - 3 - 2i)(z - 3 + 2i)$$
$$= z^2 - 3z + 2zi - 3z + 9 - 6i - 2zi + 6i - 4i^2$$
$$= z^2 - 6z + 13$$

$\therefore z^2 - 6z + 13$ must be a factor of $2z^3 + Az^2 + 32z - 13$.

Find the values of a and c:

$z^2 \times az = 2z^3 \Rightarrow a = 2$ and $(2z - c)$ is a factor.

$$(z^2 - 6z + 13)(az - c) = 2z^3 + Az^2 + 32z - 13$$

$+13 \times -c = -13 \Rightarrow c = 1$, so $(2z - 1)$ is a factor.

Find the value of A:

$$(z^2 - 6z + 13)(2z - 1) = 2z^3 - z^2 - 12z^2 + 6z + 26z - 13$$
$$= 2z^3 - 13z^2 + 32z - 13$$

$\therefore A = -13$ and the other two solutions are $\frac{1}{2}$ and $3 - 2i$.

3 Factorise and then solve $x^3 + x^2 - x + 15 = 0$.

Find the real root using the factor theorem:

Shorten your search for the right factor by considering:

- constant is 15 $\Rightarrow$ try **factors of 15: 1, 3 and 5**
- coefficients of 1 along with a constant of 15 $\Rightarrow$ **1 is not a factor**
- x^3, x^2 and 15 are all positive $\Rightarrow$ the factor must be **negative**.

$f(-5) = (-5)^3 + (-5)^2 - (-5) + 15 = -125 + 25 + 5 + 15 \neq 0$
$f(-3) = (-3)^3 + (-3)^2 - (-3) + 15 = -27 + 9 + 3 + 15 = 0 \Rightarrow (x + 3)$ is a factor.

Use long division to find $\frac{x^3 + x^2 - x + 15}{x + 3}$:

$$\begin{array}{r|l} & x^2 - 2x + 5 \\ \hline x + 3 & x^3 + x^2 - x + 15 \\ & \underline{x^3 + 3x^2} \\ & \quad -2x^2 - x \\ & \quad \underline{-2x^2 - 6x} \\ & \qquad 5x + 15 \\ & \qquad \underline{5x + 15} \\ & \qquad\qquad 0 \end{array}$$

$\therefore (x + 3)(x^2 - 2x + 5) = 0$

ISBN: 9780170447058

Use the quadratic formula to solve $x^2 - 2x + 5 = 0$:

$$x = \frac{-b \pm \sqrt{b^2 - 4ac}}{2a}$$
$$= \frac{-(-2) \pm \sqrt{(-2)^2 - 4 \times 1 \times 5}}{2 \times 1}$$
$$= \frac{2 \pm \sqrt{4 - 20}}{2}$$
$$= \frac{2 \pm \sqrt{-16}}{2}$$
$$= 1 \pm 2i$$

Factorisation: $x^3 + x^2 - x + 15 = (x + 3)(x - 1 - 2i)(x - 1 + 2i)$
Solutions: -3, $1 + 2i$ and $1 - 2i$.

Answer the following.

1 $z = -1 - 2i$ is one solution to an equation $z^3 + Az^2 - 3z - 20 = 0$.
Write down the other two solutions and find the value of A.

2 One solution of the equation $z^3 - 4z^2 + Az + 26 = 0$ is $z = 3 + 2i$.
If A is a real number, find the value of A and the other two solutions of the equation.

ISBN: 9780170447058

3 One solution of the equation $z^3 + Az^2 + 52z - 40 = 0$ is $z = 6 - 2i$.
If A is a real number, find the value of A and the other two solutions of the equation.

4 One solution of the equation $z^3 + 6z^2 + Az - 50 = 0$ is $z = -4 - 3i$.
If A is a real number, find the value of A and the other two solutions of the equation.

5 One solution of the equation $3z^3 - 11z^2 + Az + 5 = 0$ is $z = 2 - i$.
If A is a real number, find the value of A and the other two solutions of the equation.

ISBN: 9780170447058

6 One solution of the equation $5z^3 + Az^2 + 21z + 5 = 0$ is $z = 2 + i$.
If A is a real number, find the value of A and the other two solutions of the equation.

7 One solution of the equation $2z^3 - 7z^2 + Az - 15 = 0$ is $z = 1 - 2i$.
If A is a real number, find the value of A and the other two solutions of the equation.

Solve the following.

8 $z^3 + 2z^2 + 21z - 58 = 0$

9 $z^3 - 9z^2 + 19z + 29 = 0$

10 $z^3 - z^2 + 2 = 0$

11 $z^3 - 8 = 0$

ISBN: 9780170447058

Polar coordinates

- We have already seen that complex numbers can be shown on an **Argand diagram**:
- A complex number written as $\boldsymbol{z = x + yi}$ is said to be in **Cartesian** or **rectangular** form.

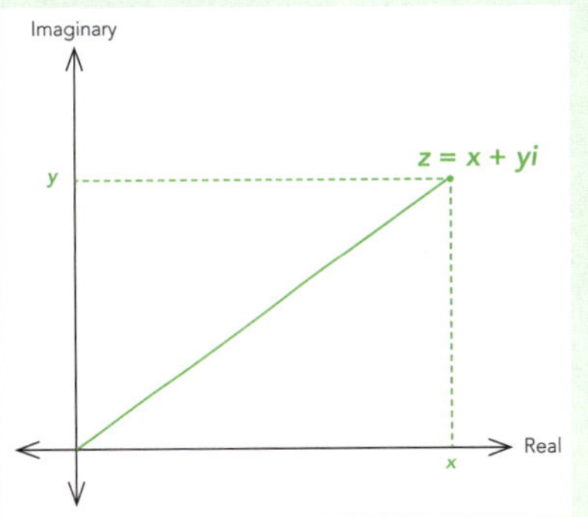

- Complex numbers can also be written in **polar form**:

The angle (θ) made with respect to the positive x-axis is known as the **argument** of z, or **arg(z)**.

Imaginary
$z = x + yi$
$r = |z|$
y
θ
Real
x

Remember: $|z|$ is also known as the **modulus**, or **mod(z)**.

You already know that $r = |z|$

$= \sqrt{x^2 + y^2}$

To find θ: $\theta = \tan^{-1} \dfrac{y}{x}$

$z = x + yi$

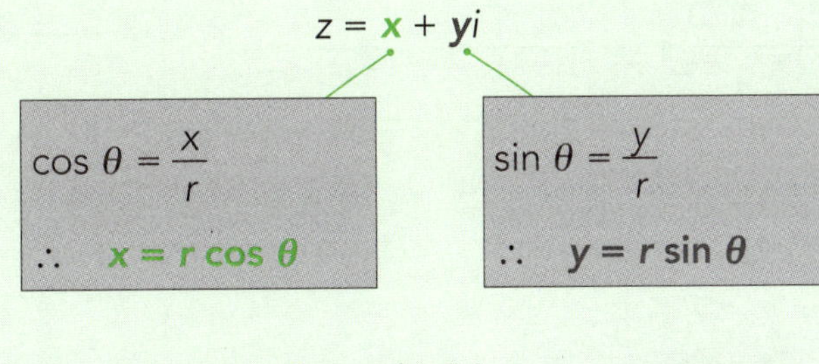

$\cos \theta = \dfrac{x}{r}$

$\therefore \quad \boldsymbol{x = r \cos \theta}$

$\sin \theta = \dfrac{y}{r}$

$\therefore \quad \boldsymbol{y = r \sin \theta}$

These are on your formula sheet.

$\therefore \quad z = r \cos \theta + r \sin \theta \, i$

$= r(\cos \theta + i \sin \theta)$

$\boldsymbol{z = r \text{ cis } \theta}$

This is **polar form**.

$r(\cos \theta + i \sin \theta)$ is shortened to $\boldsymbol{r \text{ cis } \theta}$.

Conversion from Cartesian form to polar form

Examples:

1 Write the complex number $z = 2 + 3i$ in polar form.

It's a really good idea to sketch an Argand diagram first.

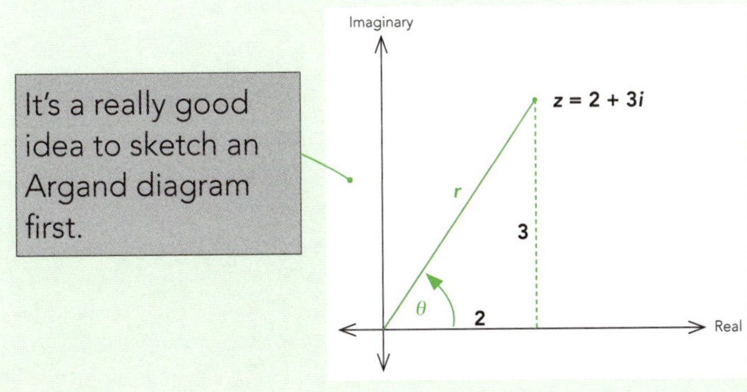

$r = \sqrt{2^2 + 3^2}$

$= \sqrt{13}$

$\theta = \tan^{-1} \dfrac{3}{2}$

$= 0.9828 \text{ or } 56.3°$

$\therefore z = \sqrt{13} \text{ cis } 0.9828$

$\text{or } z = \sqrt{13} \text{ cis } 56.3°$

ISBN: 9780170447058

2 Write the complex number $z = -\sqrt{3} - i$ in polar form.

If $\text{Re}(z) < 0$, then it is easiest to calculate α, and then $\theta = \pi - \alpha$.

$$r = \sqrt{(\sqrt{3})^2 + 1^2}$$
$$= 2$$

If you are familiar with the special triangles on page 18, you will not need to do these calculations.

$$\alpha = \tan^{-1}\frac{1}{\sqrt{3}}$$
$$= 0.5236 \text{ or } \frac{\pi}{6} \text{ or } 30°$$
$$\therefore \theta = 2.618 \text{ or } \frac{5\pi}{6} \text{ or } 150°$$

$$\therefore z = 2 \text{ cis } \frac{5\pi}{6} \text{ or } z = 2 \text{ cis } 150°$$

Convert the following complex numbers to polar form. Write your answers in degrees and in radians.

1 $z = 1 + i$

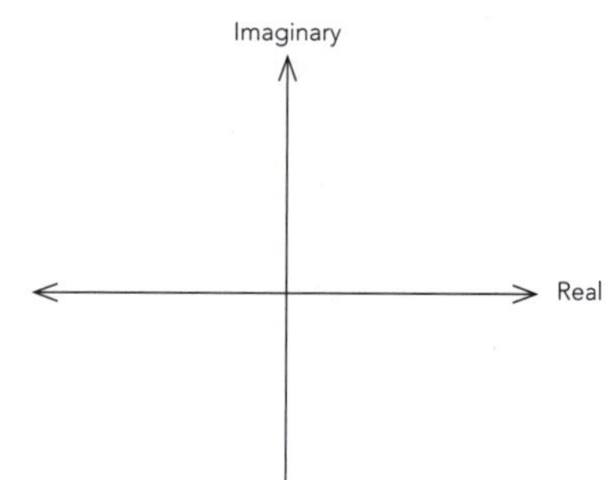

2 $z = -1 - i$

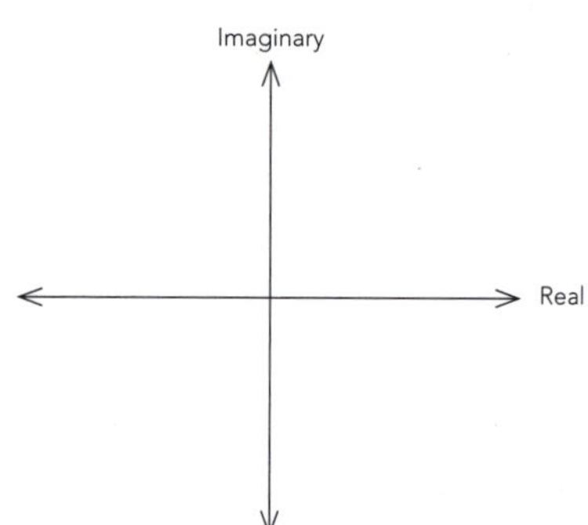

3 $z = -4 + 3i$

ISBN: 9780170447058

4 $z = 1 - \sqrt{3}i$

5 $z = 7 - 5i$

6 $z = -4 - 5i$

7 $z = 2 - i$

8 $z = 3 - 10i$

ISBN: 9780170447058

Conversion from polar form to Cartesian form

Examples:

1 Write the complex number $z = 10 \text{ cis } 0.5$ in Cartesian form.

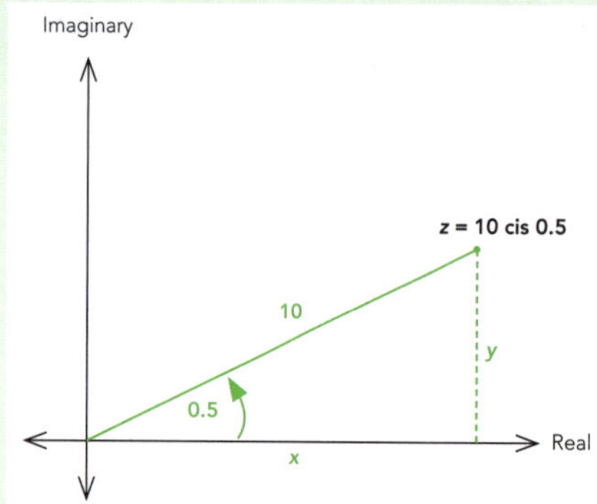

$x = r\cos\theta$

$= 10\cos 0.5$

$= 8.776$

$y = r\sin\theta$

$= 10\sin 0.5$

$= 4.794$

$\therefore z = 8.776 + 4.794\,i$

2 Write the complex number $z = \sqrt{50} \text{ cis } \left(-\frac{3\pi}{4}\right)$ in Cartesian form.

Again, if Re(z) < 0, then it is easiest to find α, and then use it to calculate x and y.

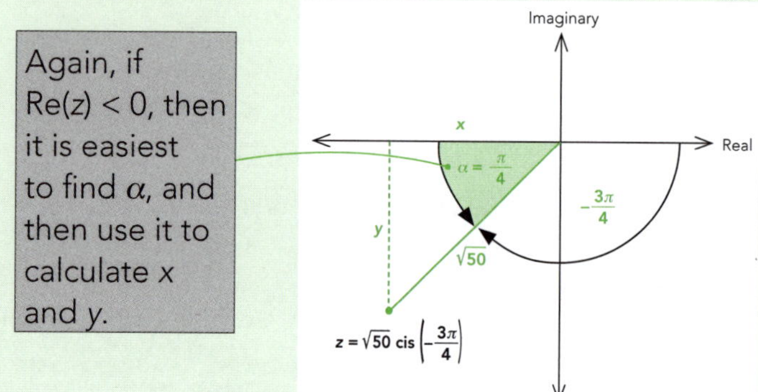

$x = r\cos\theta$

$= \sqrt{50}\cos\frac{\pi}{4}$

$= 5$

$y = r\sin\theta$

$= \sqrt{50}\sin\frac{\pi}{4}$

$= 5$

$\therefore z = -5 - 5i$

Convert the following complex numbers to Cartesian form.

9 $z = 5 \text{ cis } 0.9273$

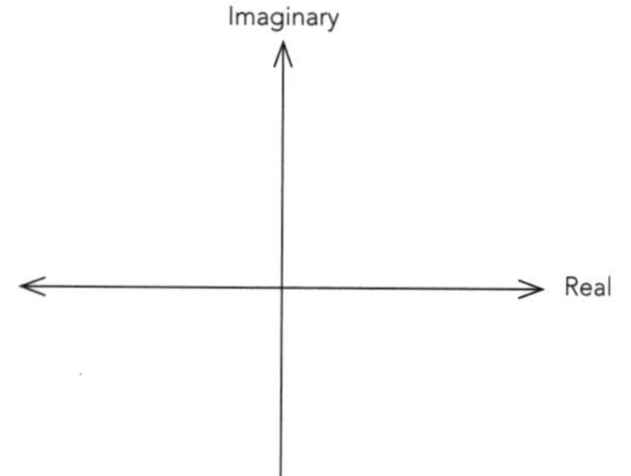

10 $z = \sqrt{2} \text{ cis } \frac{3\pi}{4}$

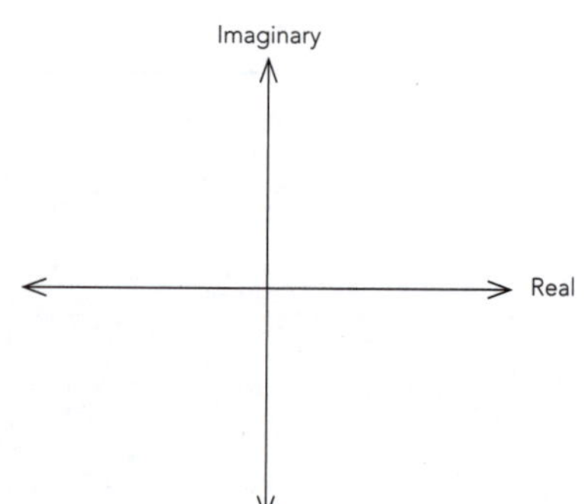

ISBN: 9780170447058

11 $z = 2 \text{ cis}\left(-\frac{2\pi}{3}\right)$

12 $z = \sqrt{101} \text{ cis } (-1.471)$

13 $z = \sqrt{18} \text{ cis } \frac{3\pi}{4}$

14 $z = 5 \text{ cis } 0.9272$

15 $z = 11.18 \text{ cis } 1.751$

ISBN: 9780170447058

Conversion using a calculator

- Some scientific calculators can do this.
- Before you start, check that your calculator is in radians (or degrees).
- These are the instructions for a graphics calculator:

→ **MENU**
→ **Run**
→ **OPTN (Options)**
→ **F6 (▷)**
→ **F5 (Angle)**
→ **F6 (▷)**

Cartesian → Polar

Example: $z = 2 + 3i$

→ **F1 (Pol()**
→ **2**
→ **,**
→ **3**
→ **)** **Exe**

You should get: **1 3.6055**

2 0.9827

So **z = 3.6055 cis 0.9827 i**.

This agrees closely with the conversion in example **1** on page 81: $z = \sqrt{13}$ cis 0.9828 ($\sqrt{13} = 3.6055$).

Polar → Cartesian

Example: $z = 10$ cis 0.5

→ **F2 (Rec()**
→ **10**
→ **,**
→ **0.5**
→ **)** **Exe**

You should get: **1 8.7750**

2 4.7942

So **z = 8.7750 + 4.7942 i**.

This agrees closely with the conversion in example **1** on page 84: $z = 8.776 + 4.794\,i$.

 ISBN: 9780170447058

Multiplying and dividing complex numbers in polar form

Multiplying: Let $z_1 = r_1 \text{ cis } \theta_1$ and $z_2 = r_2 \text{ cis } \theta_2$:

$$\mathbf{z_1 z_2 = r_1 r_2 \text{ cis } (\theta_1 + \theta_2)}$$

This is **not** on your formula sheet, so you will need to learn it!

Examples:

1 If $u = 10 \text{ cis } 0.5$ and $v = 2 \text{ cis } 1$, write the value of uv in polar form.

$$\begin{aligned} uv &= r_1 r_2 \text{ cis } (\theta_1 + \theta_2) \\ &= (10 \times 2) \text{ cis } (0.5 + 1) \\ &= 20 \text{ cis } 1.5 \end{aligned}$$

2 If $u = 5 \text{ cis } \frac{\pi}{2}$ and $v = 4 \text{ cis } \frac{\pi}{3}$, write the **exact** value of uv in polar form.

$$\begin{aligned} uv &= r_1 r_2 \text{ cis } (\theta_1 + \theta_2) \\ &= (5 \times 4) \text{ cis } (\frac{\pi}{2} + \frac{\pi}{3}) \\ &= 20 \text{ cis } \frac{5\pi}{6} \end{aligned}$$

Exact value ⇒ leave π in your answer.

Dividing: Let $z_1 = r_1 \text{ cis } \theta_1$ and $z_2 = r_2 \text{ cis } \theta_2$:

Provided $z_2 \neq 0$.

$$\mathbf{\frac{z_1}{z_2} = \frac{r_1}{r_2} \text{ cis } (\theta_1 - \theta_2)}$$

Nor is this on your formula sheet!

Examples:

1 If $u = 10 \text{ cis } 0.5$ and $v = 2 \text{ cis } 1$, write the value of $\frac{u}{v}$ in polar form.

$$\begin{aligned} \frac{u}{v} &= \frac{r_1}{r_2} \text{ cis } (\theta_1 - \theta_2) \\ &= \frac{10}{2} \text{ cis } (0.5 - 1) \\ &= 5 \text{ cis } (-0.5) \end{aligned}$$

2 If $u = 5 \text{ cis } \frac{\pi}{2}$ and $v = 4 \text{ cis } \frac{\pi}{3}$, write the **exact** value of $\frac{u}{v}$ in polar form.

$$\begin{aligned} \frac{u}{v} &= \frac{r_1}{r_2} \text{ cis } (\theta_1 - \theta_2) \\ &= \frac{5}{4} \text{ cis } \left(\frac{\pi}{2} - \frac{\pi}{3}\right) \\ &= 1.25 \text{ cis } \frac{\pi}{6} \end{aligned}$$

Once again, **exact** value ⇒ leave π in your answer.

ISBN: 9780170447058

Write answers to the following in polar form.

1 If $u = 12 \text{ cis } 0.6$ and $v = 3 \text{ cis } 0.8$, write the value of:

a uv

b $\frac{u}{v}$

2 If $u = 10 \text{ cis } \frac{2\pi}{3}$ and $v = 2 \text{ cis } \frac{\pi}{12}$, write the *exact* value of:

a uv

b $\frac{u}{v}$

3 If $u = 6 \text{ cis } 0.225$ and $v = 8 \text{ cis } 0.412$, write the value of:

a uv

b $\frac{u}{v}$

4 If $u = 3p \text{ cis } q$ and $v = p \text{ cis } 2q$, write the value of:

a uv

b $\frac{u}{v}$

ISBN: 9780170447058

Mixing it up

1 If $u = a + bi$, show that $u\bar{u} = |u|^2$.

2 z and w are complex numbers such that $z = 1 - 5i$ and $zw = 2 + 3i$. Find the exact value of $\arg(w)$.

3 z and w are complex numbers such that $w = 6 - 3i$ and $z = 1 + 2\text{i}$. Find $\frac{w}{z}$, and write your answer in polar form.

4 Write the complex number represented by the expression $w = \frac{10}{i^5}$ in polar form.

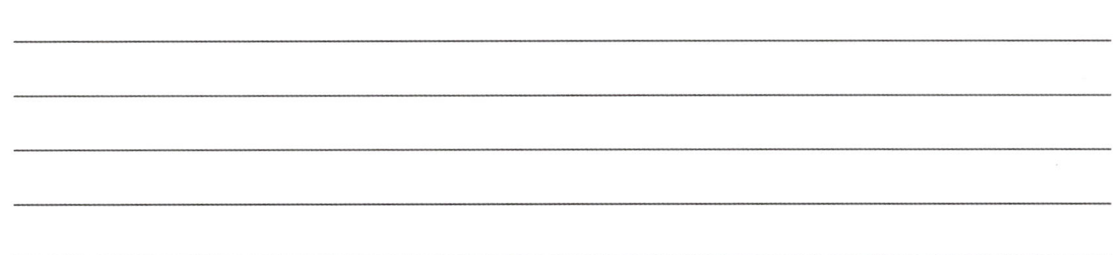

ISBN: 9780170447058

5 **a** If $u = p^3 \text{ cis } \frac{2\pi}{5}$ and $v = p \text{ cis } \frac{\pi}{4}$:
Write uv in polar form.

b Write $\frac{u}{v}$ in polar form.

c Write u^2 in polar form.

6 Write the complex number represented by $u = \left(\frac{1+i}{\sqrt{2}}\right)^2$ in Cartesian form.

7 Find the complex number w in the form of $x + yi$ if $|w\overline{w}| = 6$ and $\arg(w) = \frac{\pi}{4}$.

ISBN: 9780170447058

8 Find the complex number w in the form of $x + yi$ if $\frac{w\overline{w}}{4} = 2$ and $\arg(w) = \frac{3\pi}{4}$.

9 Let $w = a + bi$, where $w\overline{w} + 4iw = 1 + 8i$. Find all possible values for w.

10 Let $w = a + bi$, where $3w\overline{w} - 2w = 13 - 4i$. Find all possible values for w.

11 If $w = -1 + 2i$ and $z = w + \frac{3}{w}$, write z in Cartesian form, and find the value for $\arg(z)$.

ISBN: 9780170447058

12 **a** The complex number z is given by $z = \frac{1 + 2i}{p + qi}$, where p and q are real and $p > q > 0$.

Given that $\arg(z) = \frac{\pi}{4}$, show that $p = 3q$.

b Give an example of the values p and q can take, and use them calculate the value of the complex number z, writing it in rectangular form.

13 Write the complex number $\frac{i^8 + 2i^5}{4i^6 + 3i^{11}}$ in polar form.

14 If $w = (1 + \sqrt{3}i)$, write w^3 in polar form.

ISBN: 9780170447058

Powers of complex numbers in polar form: de Moivre's theorem

De Moivre's theorem states:

$$(r \text{ cis } \theta)^n = r^n \text{ cis } n\theta$$

This is on your formula sheet.

Examples:

1 Simplify $(5 \text{ cis } \frac{\pi}{4})^3$.

$(5 \text{ cis } \frac{\pi}{4})^3 = 5^3 \text{ cis } (3 \times \frac{\pi}{4})$

$= 125 \text{ cis } \frac{3\pi}{4}$

2 Simplify $(2 \text{ cis } 0.8)^5$.

$(2 \text{ cis } 0.8)^5 = 2^5 \text{ cis } (5 \times 0.8)$

$= 32 \text{ cis } 4$

$= 32 \text{ cis } (-2.283)$

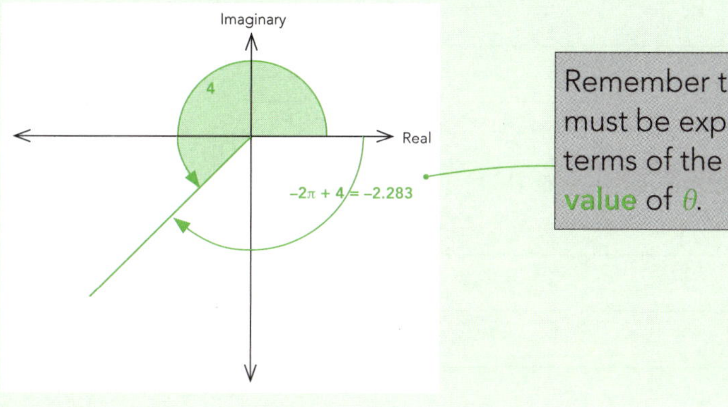

Remember that angles must be expressed in terms of the **principal value** of θ.

Simplify the following.

1 $(4 \text{ cis } 0.2)^3$

2 $(2 \text{ cis } \frac{\pi}{5})^4$

3 $\left(3 \text{ cis } \left(-\frac{\pi}{4}\right)\right)^3$

4 $\left(2 \text{ cis } \left(-\frac{\pi}{4}\right)\right)^5$

ISBN: 9780170447058

5 $\left(\sqrt{2}\ \text{cis}\left(-\frac{\pi}{3}\right)\right)^4$

6 $(0.5\ \text{cis}\ 2.5)^3$

7 If $w = 2\ \text{cis}\left(-\frac{\pi}{3}\right)$, find w^4 in the form $a + bi$, where a and b are real.

8 If $v = -3 + 3i$, find v^6 in the form $r\ \text{cis}\ \theta$.

9 If $u = 5\ \text{cis}\ 2$, find u^3 in the form $a + bi$, where a and b are real.

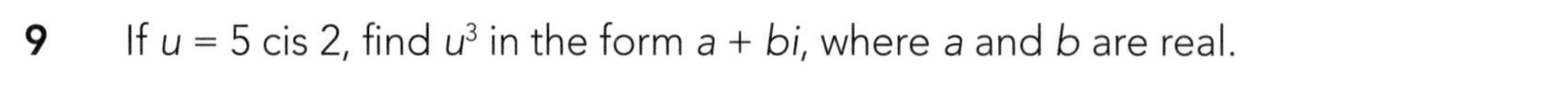

 ISBN: 9780170447058

10 An investigation.
Consider the diagram on the right.
Write z in polar form.

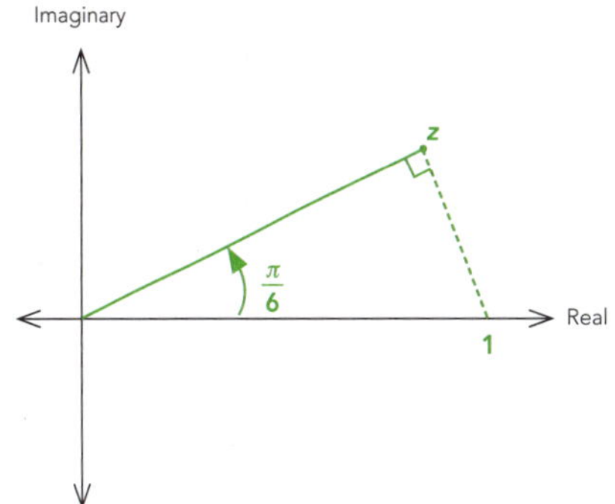

Write the values of z^2, z^3, z^4, z^5 and z^6 in polar form.

The axes below have lines through the origin at intervals of $\frac{\pi}{6}$.

At a suitable position, draw and label the complex number z^0.

Now plot approximate positions for z, z^2, z^3, z^4, z^5 and z^6.

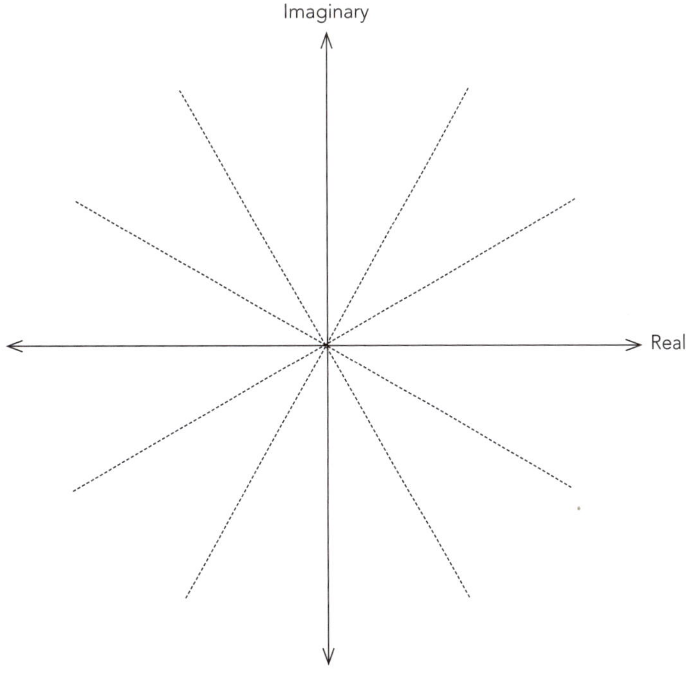

Add the complex numbers z^9 and z^{12} to the diagram.
Join the points in order with a curved line.
What shape have you drawn?

ISBN: 9780170447058

Solving polynomial equations using de Moivre's theorem

- Using de Moivre's theorem requires that equations are in **polar form**.
- Equations such as $z^3 = 2i$, $z^5 = 1$, $z^4 = -3 + 2i$ must **first** be converted into **polar form**.
- It is important to remember the fundamental theorem of algebra: when solving an equation $z^n = \ldots\ldots$, there will be **n solutions**.
- Notice that these angles are the **same**:

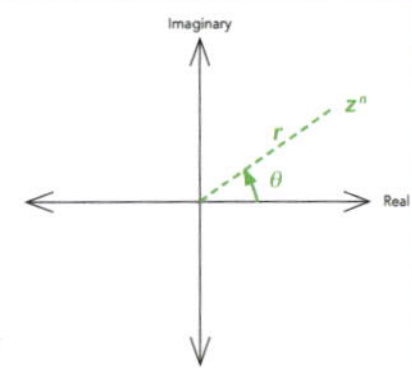

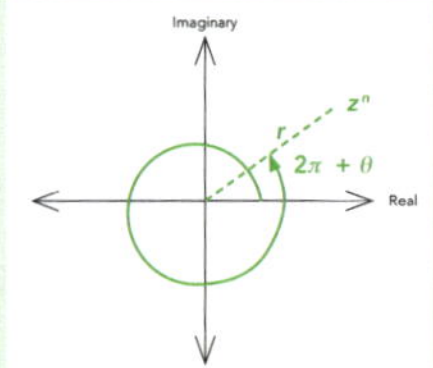

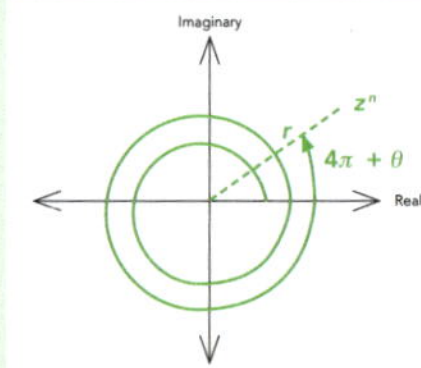

So, $\theta = 2\pi + \theta = 4\pi + \theta = 6\pi + \theta$, **etc.**

$\therefore$ $\theta = 2k\pi + \theta$, **where k is whole numbers,** $0 \leq k \leq n - 1$

Examples:

1 **Solve** $z^4 = 81 \text{ cis } \frac{\pi}{3}$ **(or 40.5 + 70.15 *i*).**

$$\text{Let } z^4 = 81 \text{ cis}\left(2k\pi + \frac{\pi}{3}\right) \Rightarrow z = (81)^{\frac{1}{4}} \text{ cis}\left(\frac{2k\pi}{4} + \frac{\pi}{12}\right)$$

You need to:

1. Substitute k = 0, 1, 2 and 3 because in this case we have z^4 so we need to find **4** roots.
2. Apply de Moivre's theorem to each expression.

To find the first root, let k = 0: $z_1 = (81)^{\frac{1}{4}} \text{ cis}\left(\frac{0}{4} + \frac{\pi}{12}\right)$

$$\therefore z_1 = 3 \text{ cis } \frac{\pi}{12}$$

To find the second root, let k = 1: $z_2 = (81)^{\frac{1}{4}} \text{ cis}\left(\frac{2\pi}{4} + \frac{\pi}{12}\right)$

$$\therefore z_2 = 3 \text{ cis } \frac{7\pi}{12}$$

To find the third root, let k = 2: $z_3 = (81)^{\frac{1}{4}} \text{ cis}\left(\frac{4\pi}{4} + \frac{\pi}{12}\right)$

$$\therefore z_3 = 3 \text{ cis } \frac{13\pi}{12} \text{ or } 3 \text{ cis}\left(-\frac{11\pi}{12}\right)$$

To find the fourth root, let k = 3: $z_4 = (81)^{\frac{1}{4}} \text{ cis}\left(\frac{6\pi}{4} + \frac{\pi}{12}\right)$

$$\therefore z_4 = 3 \text{ cis } \frac{19\pi}{12} \text{ or } 3 \text{ cis}\left(-\frac{5\pi}{12}\right)$$

Note: Substituting other values for k would produce duplicates of these roots.

ISBN: 9780170447058

So roots to $z^4 = 81 \text{ cis } \frac{\pi}{3}$ are $z = 3 \text{ cis } \frac{\pi}{12}$, $3 \text{ cis } \frac{7\pi}{12}$, $3 \text{ cis } \left(-\frac{11\pi}{12}\right)$ and $3 \text{ cis } \left(-\frac{5\pi}{12}\right)$.

Notice that these occur at **regular intervals** of θ: $\frac{6\pi}{12}$ or $\frac{\pi}{2}$.

This is because there are four roots within 2π, so the solutions are at intervals of $\frac{\pi}{2}$.

Drawing these roots:

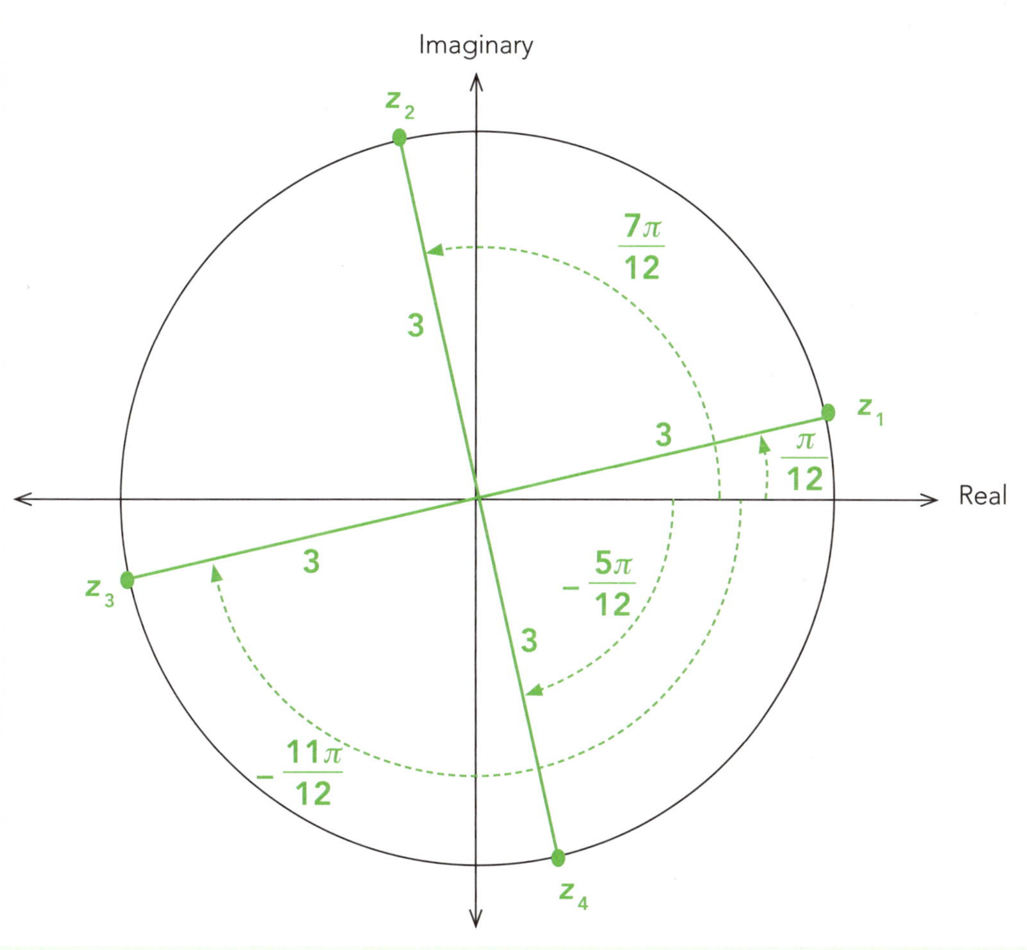

Notice that:

1 All the roots lie on the circumference of a circle with its centre at the origin and a radius of 3.

2 As we observed above, there is a right angle $\left(\frac{\pi}{2} \text{ or } \frac{6\pi}{12}\right)$ between any two adjacent roots.

ISBN: 9780170447058

2 Solve $z^3 = 125$. $125 = 125 \text{ cis } 0$. Write the complex number in **polar form**.

$$\text{Let } z^3 = 125 \text{ cis } (2k\pi + 0) = 125 \text{ cis } 2k\pi \Rightarrow z = (125)^{\frac{1}{3}} \text{ cis } \left(\frac{2k\pi}{3}\right)$$

To find the first root, let k = 0: $z_1 = (125)^{\frac{1}{3}} \text{ cis } 0$

$\therefore\ z_1 = 5 \text{ cis } 0 = 5$

To find the second root, let k = 1: $z_2 = (125)^{\frac{1}{3}} \text{ cis } \left(\frac{2\pi}{3}\right)$

$\therefore\ z_2 = 5 \text{ cis } \frac{2\pi}{3}$

To find the third root, let k = 2: $z_3 = (125)^{\frac{1}{3}} \text{ cis } \left(\frac{4\pi}{3}\right)$

$\therefore\ z_3 = 5 \text{ cis } \frac{4\pi}{3} \text{ or } 5 \text{ cis } \left(-\frac{2\pi}{3}\right)$

So roots to $z^3 = 125$ are $z = 5,\ 5 \text{ cis } \frac{2\pi}{3} \text{ or } 5 \text{ cis } \left(-\frac{2\pi}{3}\right)$

Drawing these roots:

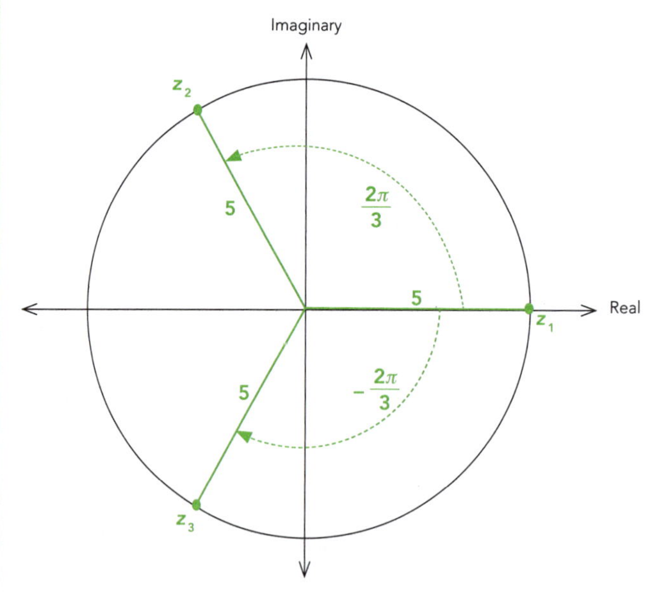

Notice that:

1 All the roots lie on the circumference of a circle with its centre at the origin and a radius of 5.

2 The angles between any two adjacent roots is $\frac{2\pi}{3}$, so once again they are spaced at equal angles around the origin.

ISBN: 9780170447058

3 Solve $z^6 = 64i$. $64i = 64 \text{ cis } \frac{\pi}{4}$.

Write the complex number in **polar form**.

$$\text{Let } z^6 = 64 \text{ cis}\left(2k\pi + \frac{\pi}{4}\right) \Rightarrow z = (64)^{\frac{1}{6}} \text{ cis}\left(\frac{2k\pi}{6} + \frac{\pi}{24}\right)$$

To find the first root, let k = 0: $z_1 = (64)^{\frac{1}{6}} \text{ cis}\left(\frac{0}{6} + \frac{\pi}{24}\right)$

$\therefore z_1 = 2 \text{ cis } \frac{\pi}{24}$

To find the second root, let k = 1: $z_2 = (64)^{\frac{1}{6}} \text{ cis}\left(\frac{2\pi}{6} + \frac{\pi}{24}\right)$

$\therefore z_2 = 2 \text{ cis } \frac{9\pi}{24}$

To find the third root, let k = 2: $z_3 = (64)^{\frac{1}{6}} \text{ cis}\left(\frac{4\pi}{6} + \frac{\pi}{24}\right)$

$\therefore z_3 = 2 \text{ cis } \frac{17\pi}{24}$

Notice the pattern here: π, 9π, 17π. These are at intervals of 8π, which makes calculation of the remaining roots unnecessary.

The fourth must be: $z_4 = 2 \text{ cis } \frac{25\pi}{24} = 2 \text{ cis}\left(-\frac{23\pi}{24}\right)$

The fifth must be: $z_5 = 2 \text{ cis } \frac{33\pi}{24} = 2 \text{ cis}\left(-\frac{15\pi}{24}\right)$

The sixth must be: $z_6 = 2 \text{ cis } \frac{41\pi}{24} = 2 \text{ cis}\left(-\frac{7\pi}{24}\right)$

Drawing these roots:

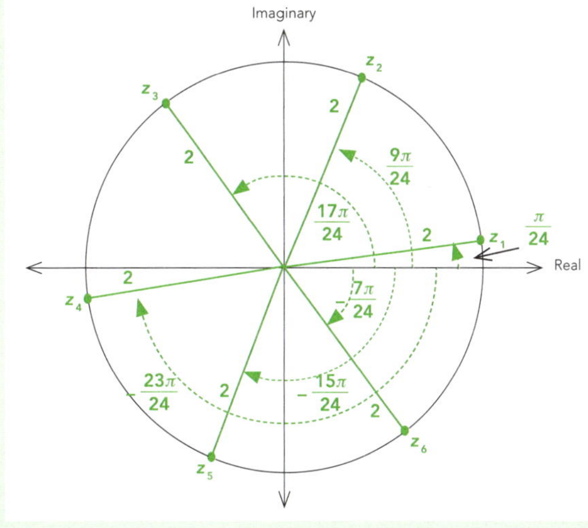

Once again, notice that:

1 All the roots lie on the circumference of a circle with its centre at the origin and a radius of 2.

2 The angles between any two adjacent roots is $\frac{\pi}{3}$ (or $\frac{8\pi}{24}$), so once again they are spaced at equal angles around the origin.

Answer the following questions, and write your answers in polar form unless the question asks otherwise.

1 Solve the equation $z^2 = 2 \text{ cis } \frac{\pi}{3}$.

2 Find each of the roots of the equation $z^3 = 8$.

3 Solve the equation $z^5 + 32 = 0$.
Hint: $-32 = 32 \text{ cis } \pi$.

ISBN: 9780170447058

4 Find all the square roots of $-i$.

5 Find each of the cube roots of the expression $-1 + i$.

6 Find $\sqrt[3]{a^3}$, where a is real.

ISBN: 9780170447058

7 Find $\sqrt[3]{-4 + 4\sqrt{3}i}$.

8 Solve the equation $z^5 = ki$, where k is real. Write your solutions in polar form in terms of k.

9 Solve the equation $z^4 = -9p^2$, where p is real. Write your solutions in polar form in terms of p.

  ISBN: 9780170447058

Finding the locus of complex numbers subject to restrictions

- The complex number $z = x + yi$ can be represented by a point (x, y) on the Argand plane.
- If restrictions are placed on z, then the set of all the resulting points is called the **locus**.

1 Where the real and/or imaginary part of the complex number is restricted

Examples: Find the equivalent Cartesian equations for the following loci and sketch each one on the Argand diagrams below. Let $z = x + yi$.

1 $\text{Re}(z) = 5$

$\text{Re}(x + yi) = 5 \Rightarrow x = 5$

Locus:

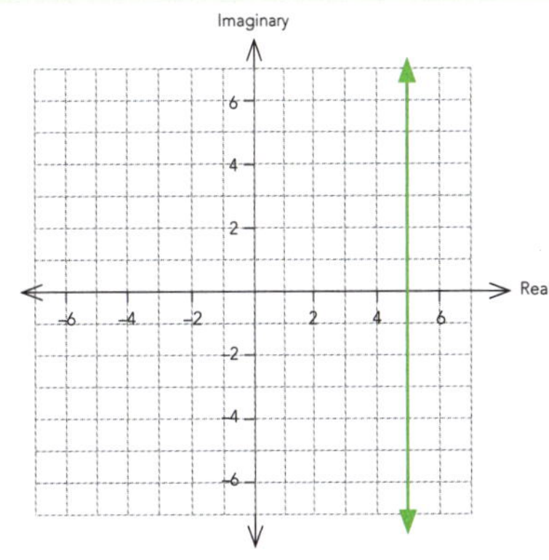

2 $\text{Im}(z) = -3$

$\text{Im}(x + yi) = -3 \Rightarrow y = -3$

Locus:

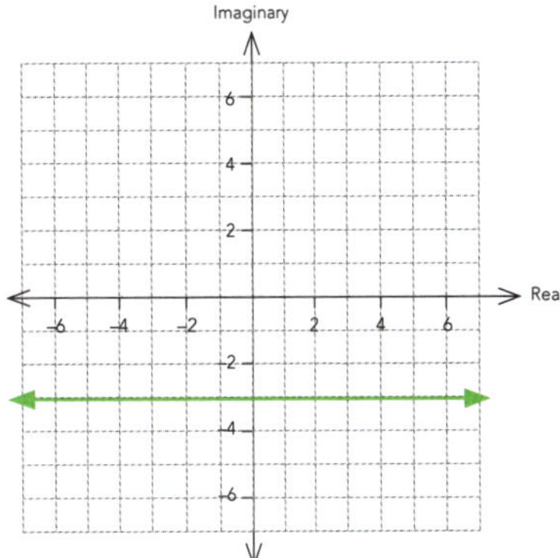

3 $3\text{Re}(z) - 2\text{Im}(z) = 6$

$3\text{Re}(x + yi) - 2\text{Im}(x + yi) = 6$

$\Rightarrow \quad 3x - 2y = 6$

Locus:

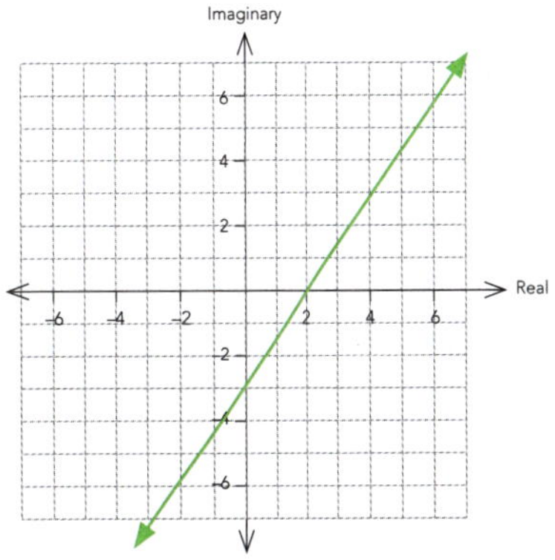

ISBN: 9780170447058

Find the equivalent Cartesian equations for the following loci and sketch each one on the Argand diagrams below. Let $z = x + yi$.

1 $\text{Im}(z) = 2$

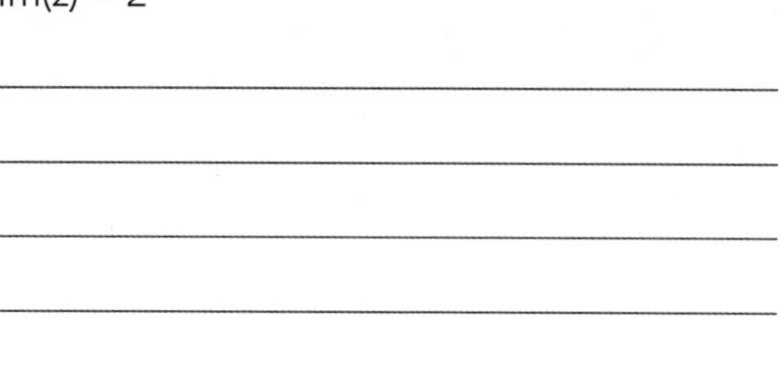

2 $\text{Re}(z) = -4$

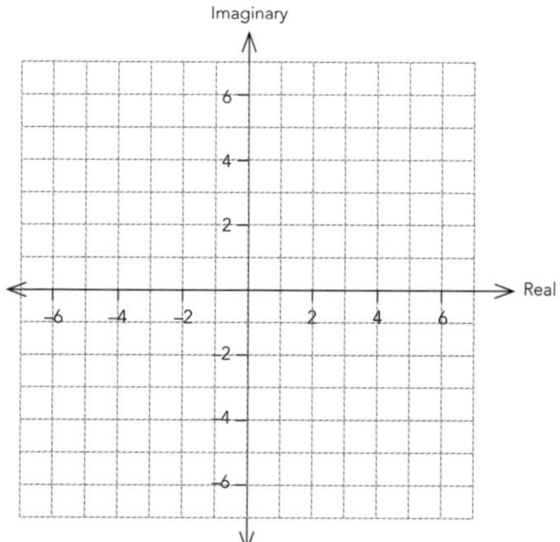

3 $3\text{Re}(z) = 12$

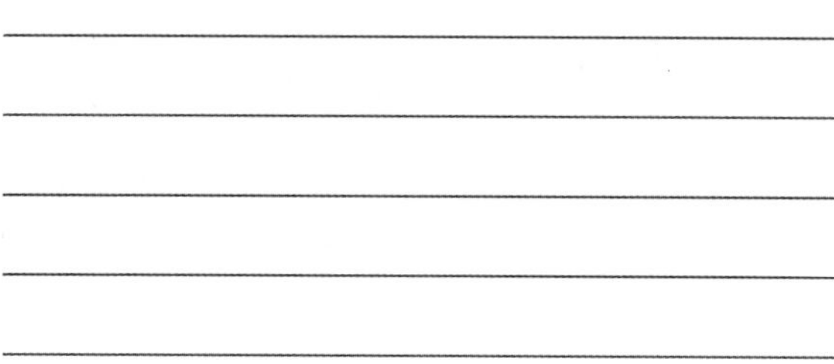

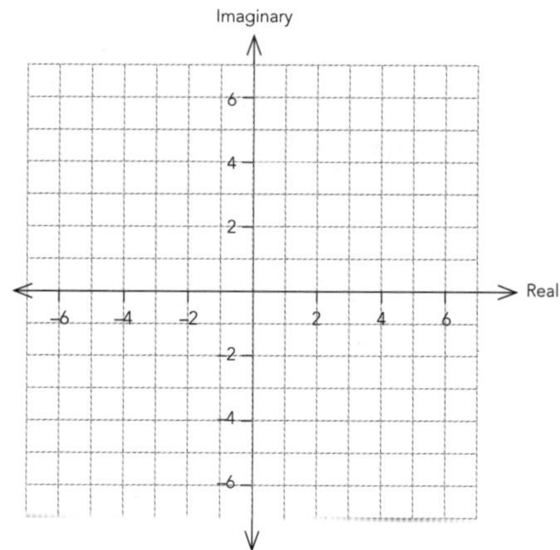

4 $\text{Re}(z) = \text{Im}(z) + 1$

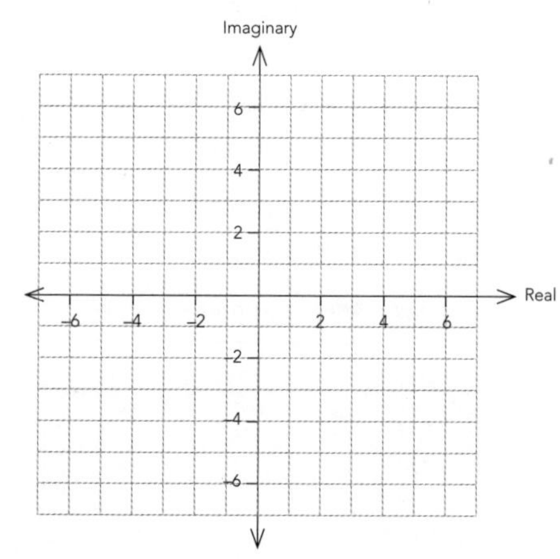

 ISBN: 9780170447058

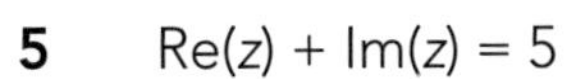

5 $\text{Re}(z) + \text{Im}(z) = 5$

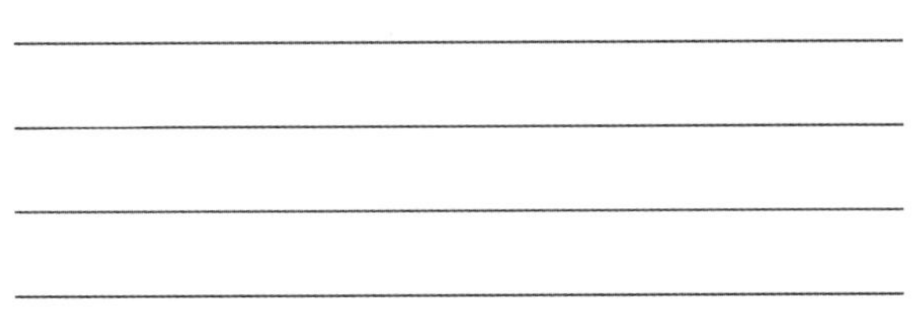

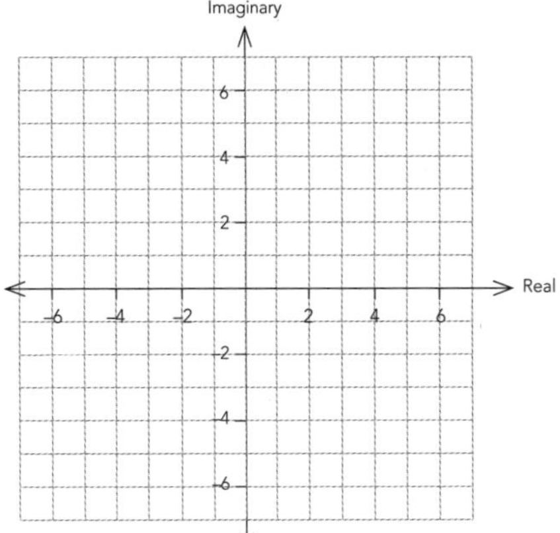

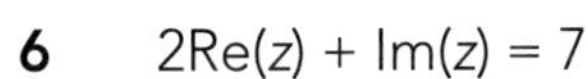

6 $2\text{Re}(z) + \text{Im}(z) = 7$

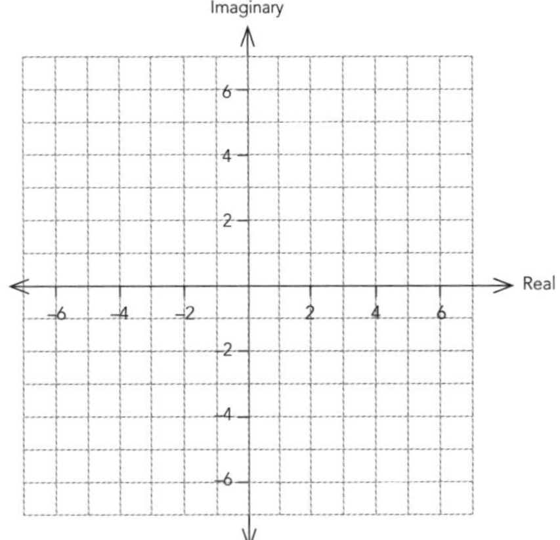

7 $3\text{Re}(z) + 2\text{Im}(z) = -12$

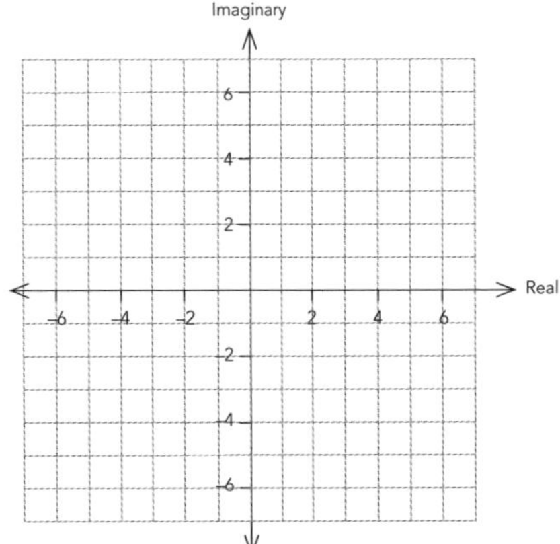

8 $\text{Re}(z) - \text{Im}(z) \le 6$

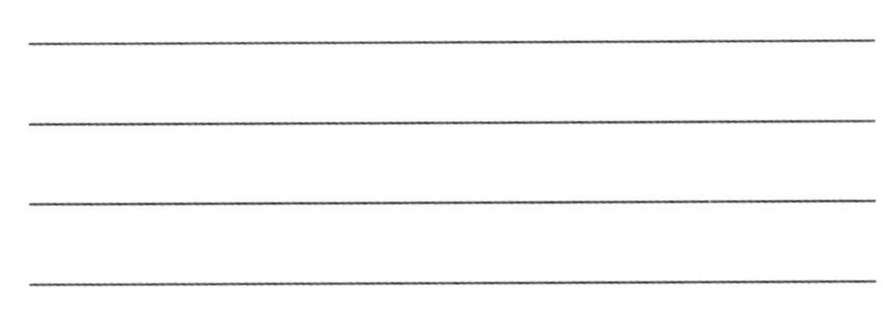

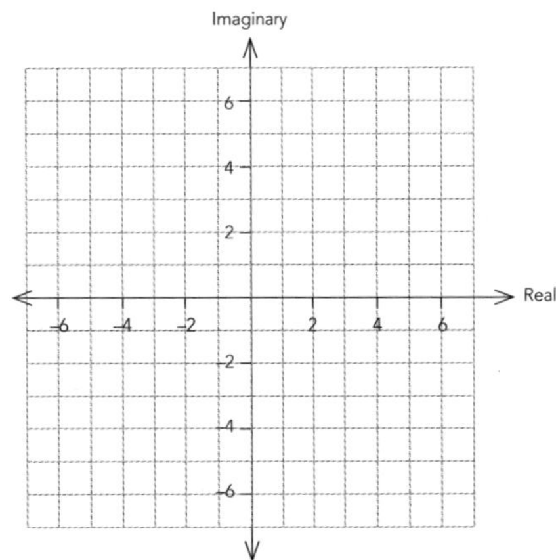

ISBN: 9780170447058

2 Where the modulus of the complex number is restricted

Examples:

1 Find the Cartesian equation of the locus described by $|z| = 3$.

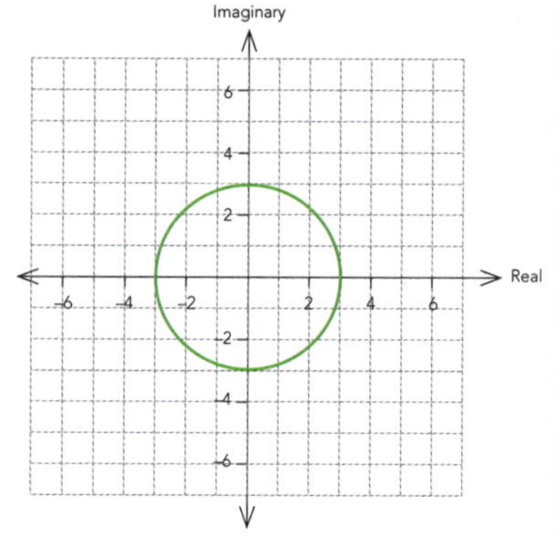

$$|(x + yi)| = 3 \Rightarrow \sqrt{(x^2 + y^2)} = 3$$

$$\therefore x^2 + y^2 = 9$$

This is the equation for a circle with its centre at (0, 0) and a radius of 3.

2 Find the Cartesian equation of the locus described by $|z + 3 - 2i| = 5$.

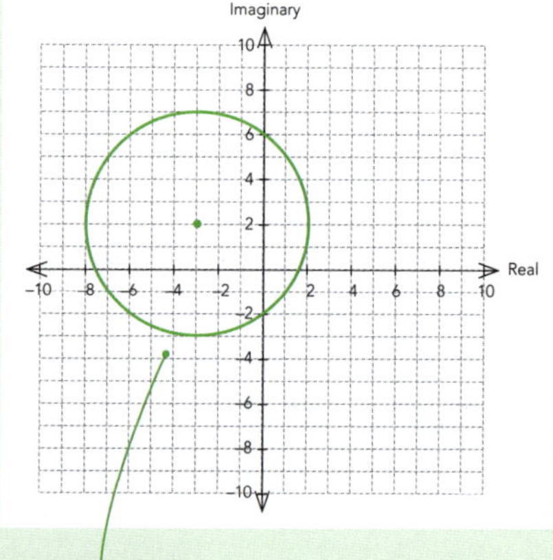

$$|z + 3 - 2i| = 5 \Rightarrow |x + yi + 3 - 2i| = 5$$

$$|(x + 3) + i(y - 2)| = 5$$

$$\sqrt{(x + 3)^2 + (y - 2)^2} = 5$$

$$\therefore (x + 3)^2 + (y - 2)^2 = 25$$

This is the equation for a circle with its centre at (–3, 2) and a radius of 5.

You can think of this as a circle with a radius of 5, which is translated 3 units to the left and 2 units up.

3 The complex number $u = 2 + bi$ is on the locus of the points defined by $|z - 4 + 2i| = |z + 3 - i|$. Find the value of b.

$$|(x - 4) + i(y + 2)| = |(x + 3) + i(y - 1)|$$

$$\sqrt{(x - 4)^2 + (y + 2)^2} = \sqrt{(x + 3)^2 + (y - 1)^2}$$

$$(x - 4)^2 + (y + 2)^2 = (x + 3)^2 + (y - 1)^2$$

$$x^2 - 8x + 16 + y^2 + 4y + 4 = x^2 + 6x + 9 + y^2 - 2y + 1$$

$$6y - 14x = -10$$

$$3y - 7x = -5$$

$$u = 2 + bi \Rightarrow x = 2$$

$$\therefore 3y - 14 = -5, \text{ so } y = b = 3$$

ISBN: 9780170447058

Answer the following questions.

1 Find the equivalent Cartesian equations for the following loci and sketch each one on the Argand diagrams below. Let $z = x + yi$.

a $|z| = 5$

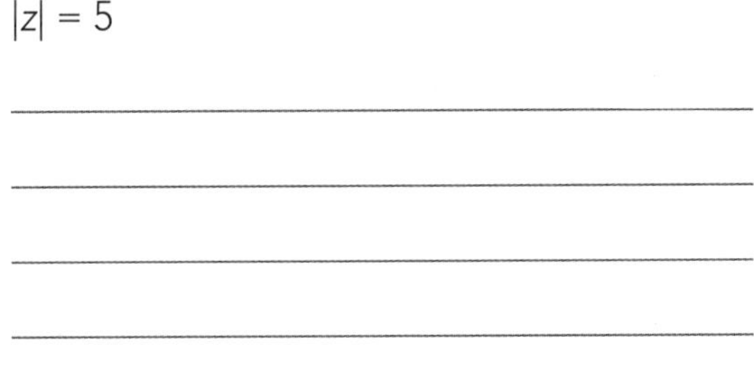

b $|z + 3i| = 4$

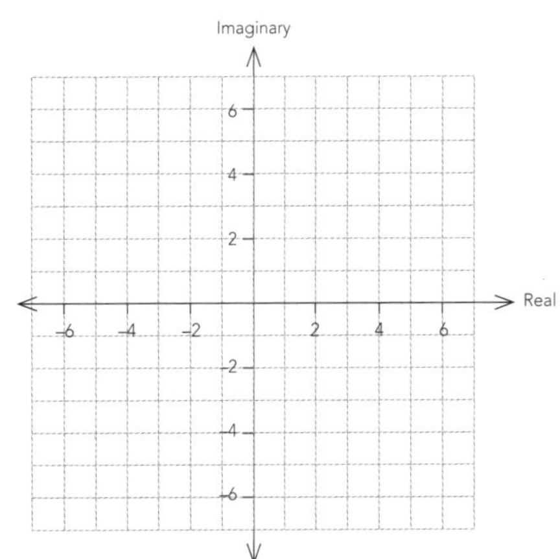

2 Find the Cartesian equation of the locus described by $|z - 5| = 3$.

3 Find the Cartesian equation of the locus described by $|z - 1 + 2i| = 7$.

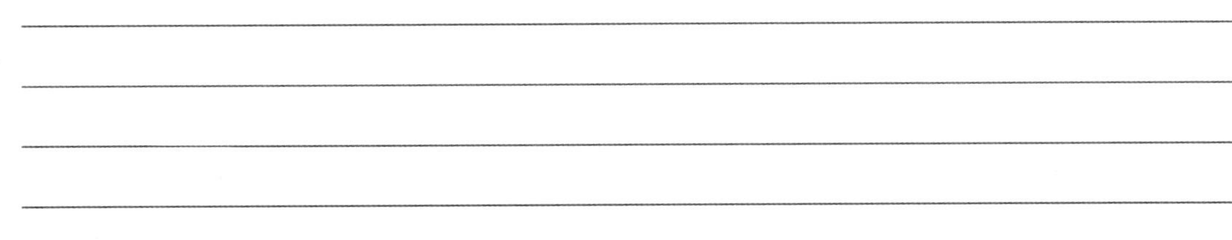

ISBN: 9780170447058

4 The locus described by $|z - 3| = |z + i|$ is a straight line. Find the gradient of the line.

5 The locus described by $|z + 2 - i| = |z - 3|$ is a straight line. Find the equation of the line.

6 The complex number $v = 2 + mi$ is on the locus of the points defined by $|z + 6 - i| = |z + 3 + 2i|$. Find the value of m.

 ISBN: 9780170447058

7 Find the Cartesian equation described by $2|z| = |z + 3 - 3i|$. Write your answer in the form $(x + A)^2 + (y + B)^2 = K$.
Hint: You will need to complete the square in order to find the answer.

8 Find the Cartesian equation described by $2|z + 1 - 3i| = |z - 2 + 6i|$. Write your answer in the form $(x + A)^2 + (y + B)^2 = K$.

ISBN: 9780170447058

3 Where the argument of the complex number is restricted

Examples: Find the Cartesian equations for the following complex loci.

1

$$\arg(z - 2 + 3i) = \frac{\pi}{4}$$

$$\arg((x - 2) + i(y + 3)) = \frac{\pi}{4}$$

$$\frac{\text{Im}(z)}{\text{Re}(z)} = \frac{y + 3}{x - 2} = \tan\frac{\pi}{4} = 1$$

$$\therefore y + 3 = x - 2$$

$$y = x - 5$$

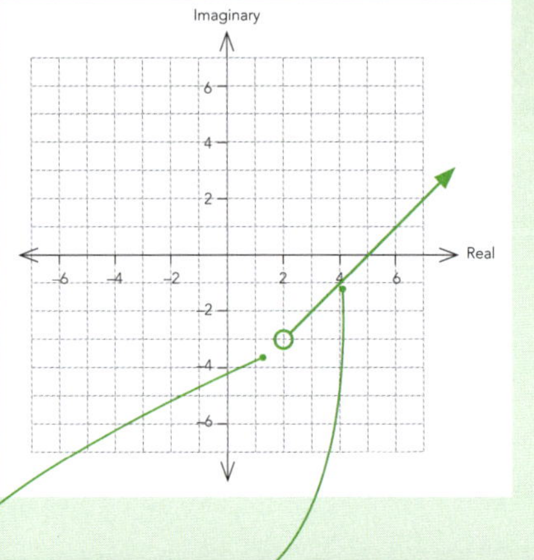

Because $\frac{\pi}{4}$ is in the first quadrant ( ), $x - 2 > 0 \Rightarrow x > 2$ and $y + 3 > 0 \Rightarrow y > -3$.

Because the denominator is $x - 2$, $x \neq 2$. Indicate this on your graph **o**, where $x = 2$.

This could be considered to be a translation of $\arg(z) = \frac{\pi}{4}$ to the **right 2** and **down 3**.

An alternative view:

- $z - 2 + 3i = (x + yi) - (2 - 3i)$, which represents a vector from the point (2, –3) to the point (x, y).
- Because $\arg(z - 2 + 3i) = \frac{\pi}{4}$, the vector must be at an angle of $\frac{\pi}{4}$ to a line through (2, –3) and parallel with the x-axis.

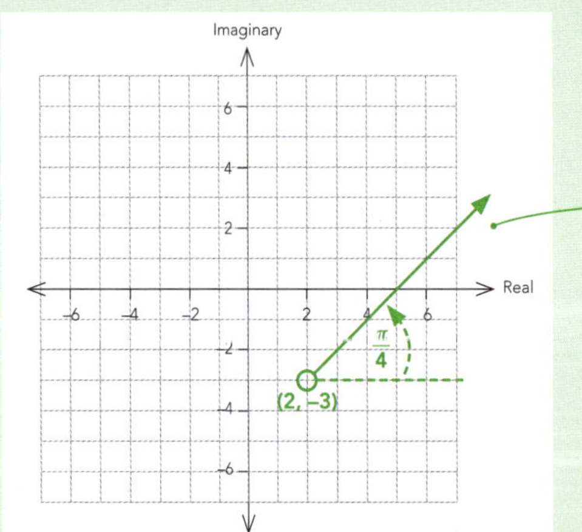

A line with a limited domain and/or range is sometimes called a **half-line**.

In general: $\arg(z - z_1) = \theta$ is represented by the half-line that
- **starts from the point z_1 and**
- **makes an angle (θ) from a line which passes through (x_1, y_1) and is parallel with the x-axis.**

ISBN: 9780170447058

2

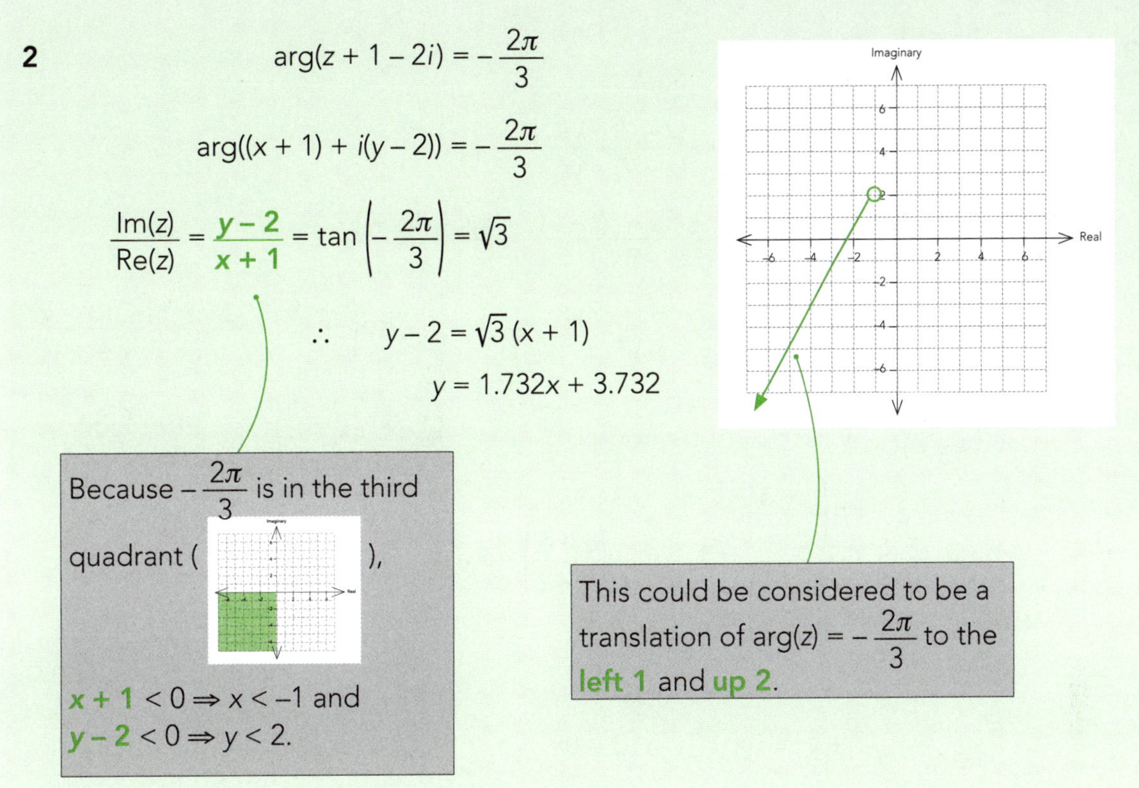

The alternative view:

- $z + 1 - 2i = (x + yi) - (-1 + 2i)$, which represents a vector from the point $(-1, 2)$ to the point (x, y).
- Because $\arg(z + 1 - 2i) = -\frac{2\pi}{3}$, the vector must be at an angle of $-\frac{2\pi}{3}$ to a line through $(-1, 2)$ and parallel with the x-axis.

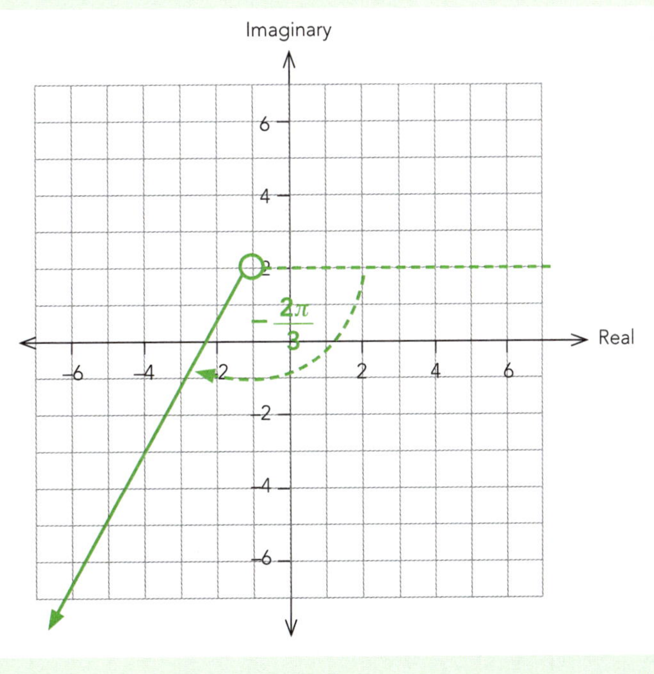

ISBN: 9780170447058

3 $\arg\left(\frac{z+5}{z-2}\right) = \frac{\pi}{4}$

$$\frac{z+5}{z-2} = \frac{(x+5)+yi}{(x-2)+yi} \times \frac{(x-2)-yi}{(x-2)-yi}$$

$$= \frac{x^2 - 2x - xyi + 5x - 10 - 5yi + xyi - 2yi + y^2}{x^2 - 4x + 4 + y^2}$$

$$= \frac{\mathbf{x^2 + 3x - 10 + y^2 - 7yi}}{x^2 - 4x + 4 + y^2}$$

$\therefore \text{Re}\left(\frac{z+5}{z-2}\right) = \frac{x^2 + 3x - 10 + y^2}{x^2 - 4x + 4 + y^2}$

$\text{Im}\left(\frac{z+5}{z-2}\right) = \frac{-7y}{x^2 - 4x + 4 + y^2}$

But $\arg\left(\frac{z+5}{z-2}\right) = \frac{\pi}{4}$

$$\tan\frac{\pi}{4} = \frac{\text{Im}\left(\frac{z+5}{z-2}\right)}{\text{Re}\left(\frac{z+5}{z-2}\right)} = 1 \Rightarrow \text{Im}\left(\frac{z+5}{z-2}\right) = \text{Re}\left(\frac{z+5}{z-2}\right)$$

$$\therefore \quad \mathbf{x^2 + y^2 + 3x - 10 = -7y}$$

$$x^2 + y^2 + 3x + 7y - 10 = 0$$

The denominators for Re(z) and Im(z) are the same, so you need to consider **only** the **numerators**.

Answer the following questions.

1 Find the Cartesian equation for the complex locus described by $\arg(z) = -\frac{\pi}{4}$. Draw your solution on the grid.

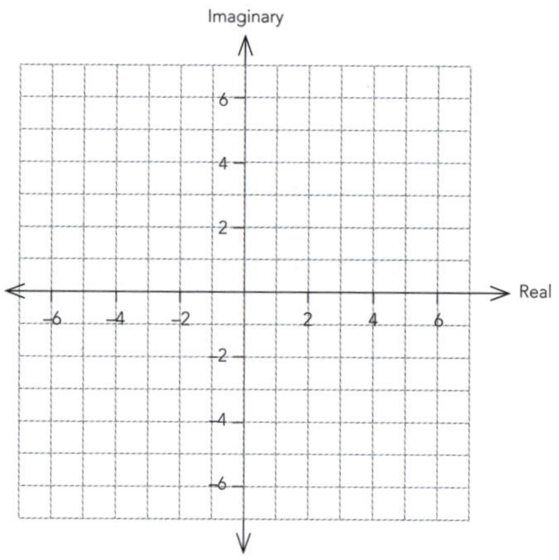

2 Find the Cartesian equation for the complex locus described by $\arg(z) = \frac{5\pi}{6}$. Draw your solution on the grid.

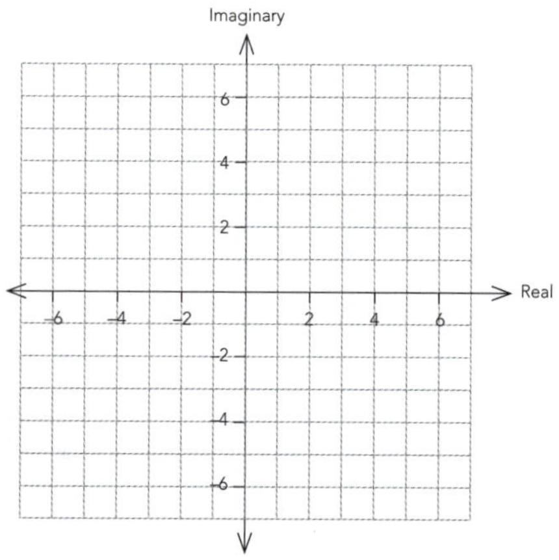

ISBN: 9780170447058

3 Find the Cartesian equation for the complex locus described by $\arg(z-2) = -\frac{\pi}{4}$. Draw your solution on the grid.

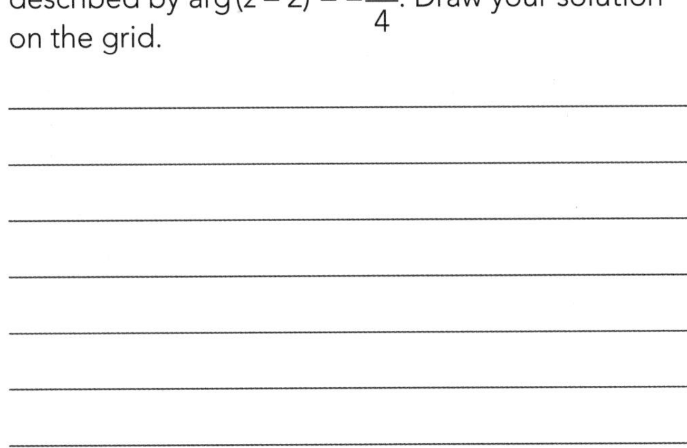

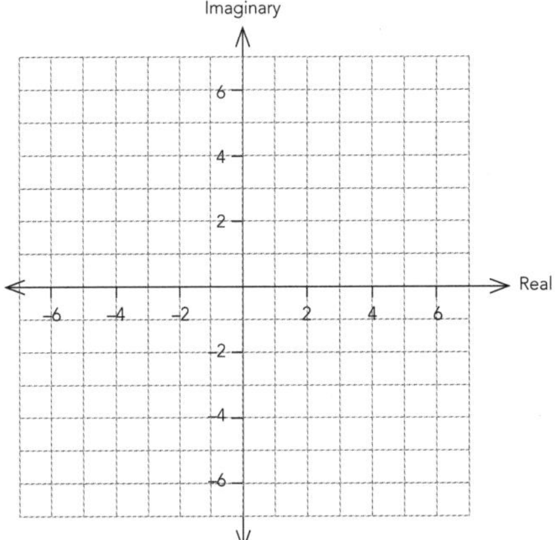

4 Find the Cartesian equation for the complex locus described by $\arg(z+i) = \frac{\pi}{3}$. Draw your solution on the grid.

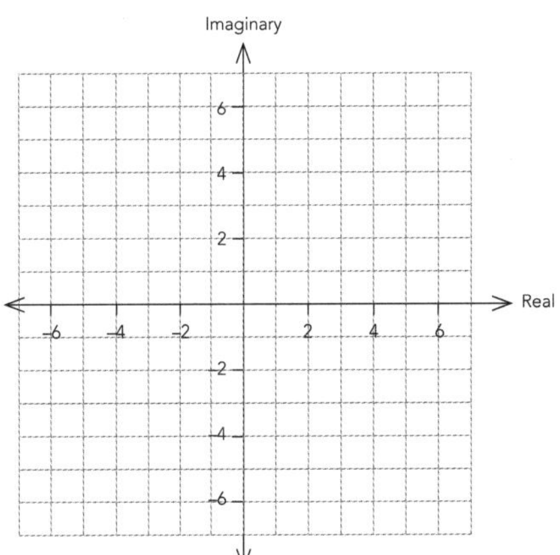

5 Find the Cartesian equation for the complex locus described by $\arg(z+2-3i) = \frac{\pi}{4}$. Draw your solution on the grid.

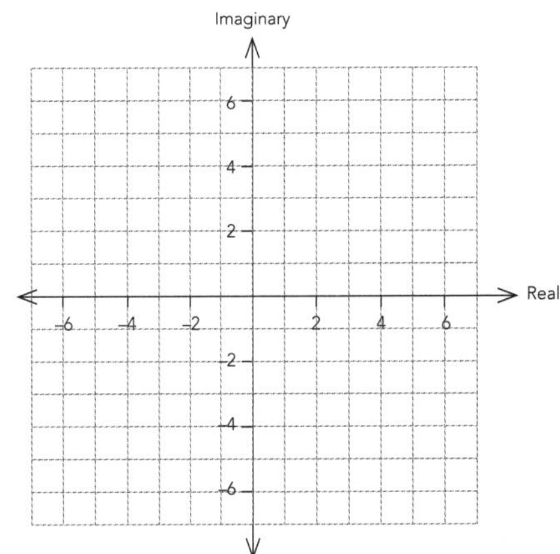

ISBN: 9780170447058

6 Find the Cartesian equation for the complex locus described by $\arg(z - 1 + 4i) = \frac{\pi}{6}$. Draw your solution on the grid.

7 Find the Cartesian equation for the complex locus described by $\arg\left(\frac{z}{z-1}\right) = \frac{\pi}{4}$.

8 Find the Cartesian equation for the complex locus described by $\arg\left(\frac{z-3}{z+1}\right) = \frac{\pi}{4}$.

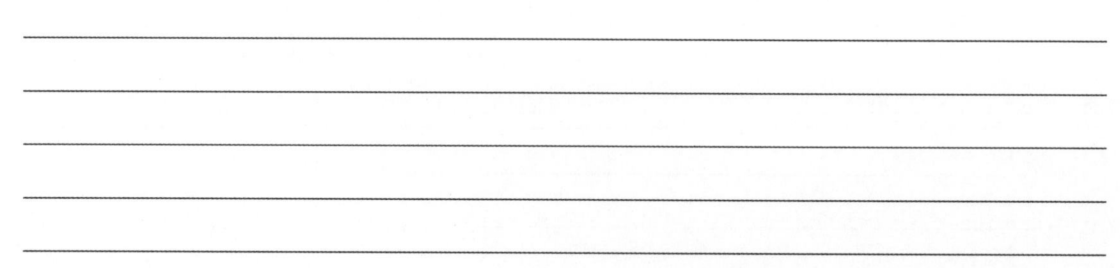

 ISBN: 9780170447058

9 Find the Cartesian equation for the complex locus described by $\arg\left(\frac{z+4}{z-1}\right) = \frac{\pi}{4}$.

10 Find the Cartesian equation for the complex locus described by $\arg\left(\frac{z-3i}{z+4}\right) = \frac{\pi}{4}$.

11 Find the Cartesian equation for the complex locus described by $\arg\left(\frac{z+2i}{z-3}\right) = \frac{\pi}{4}$.

ISBN: 9780170447058

Practice questions

Practice question one

a If $u = p^4 \text{ cis } \frac{3\pi}{4}$ and $v = p \text{ cis } \frac{\pi}{3}$, write the exact value of $\frac{u}{v}$.

b Dividing $x^3 + 5x^2 - 17x + m$ by $(x - 2)$ gives a remainder of -7. Find the value of m.

c Solve the equation $\sqrt{x} = \sqrt{2k + x} + 5$ for x in terms of k.

 ISBN: 9780170447058

d Let $w = 4 - 3i$. If $v = \dfrac{1}{|\bar{w}| - w}$, write v in the form $v = a \pm bi$.

e Solve $z^3 = \dfrac{2p}{\sqrt{2}} + \dfrac{2p}{\sqrt{2}}i$.

ISBN: 9780170447058

Practice question two

a The complex numbers u and v are represented on the Argand diagram on the right.
If $w = u - \overline{v}$, show w on the same diagram.

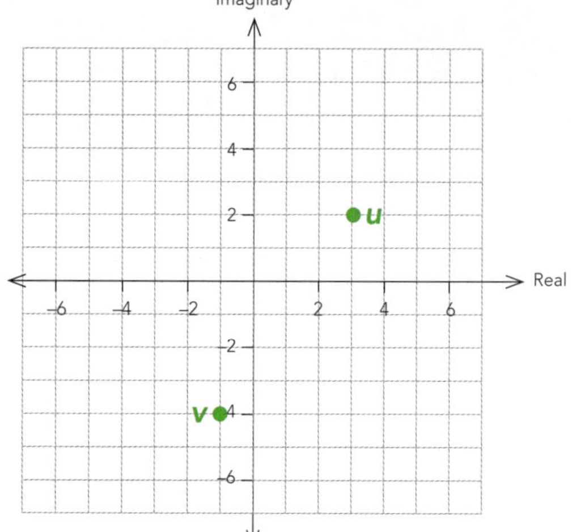

b If $z = (1 - i)^3$, find $\arg(z)$.

c Find all possible values of k that make $u = \dfrac{k - 12i}{3 - ki}$ a purely real number.

ISBN: 9780170447058

d $\frac{10x^3 + 31x^2 + x - 11}{5x - 2} = 2x^2 + Ax + Bx + \frac{C}{5x - 2}$, where A, B and C are integers.

Find the values of A, B and C.

e Find the Cartesian equation for the complex locus described by $\arg(z + 5i - 2) = \frac{3\pi}{4}$.

Draw your solution on the grid.

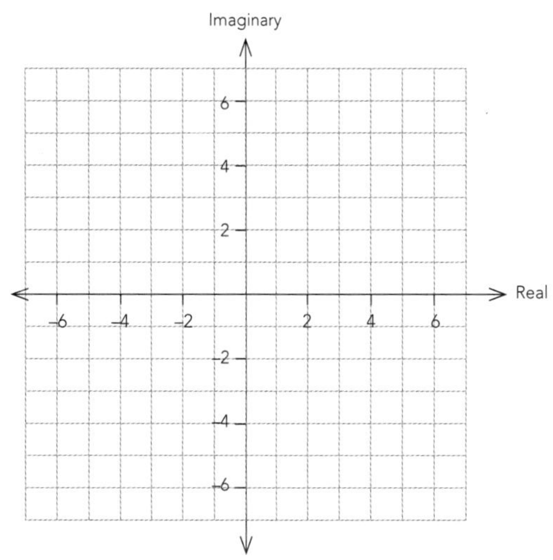

ISBN: 9780170447058

Practice question three

a Write $\frac{5}{4 + \sqrt{6}}$ in the form $a + b\sqrt{6}$.

b If $u = -5 + 5i$, find u^4 in terms of $r \operatorname{cis} \theta$.

c One solution to the equation $z^3 - z^2 + Az + 65 = 0$ is $z = 3 + 2i$.
If A is a real number, find the value of A, and the other two solutions to the equation.

 ISBN: 9780170447058

d For what values of p, where p is real, does the graph of $y = 2px^2 - 6px + 5$ not intersect the x-axis?

e Find the Cartesian equation described by $2|z - 1| = |z + 2 - 3i|$. Write your answer in the form $(x + A)^2 + (y + B)^2 = K$, and describe the locus represented by this equation.

Answers

Revision and useful background (pp. 7–22)

1 Polynomial equations (pp. 7–11)

Expanding three brackets

1 $x^3 - 13x + 12$
2 $2x^3 + 3x^2 - 32x + 15$
3 $6y^3 + 13y^2 + 4y - 3$
4 $2p^3 + 13p^2 + 8p - 48$
5 $64y^3 - 240y^2 + 300y - 125$
6 $x^6 + 9x^4y + 27x^2y^2 + 27y^3$

Solving using the quadratic formula

1 $-2 \pm \frac{\sqrt{40}}{2}$ or $-2 \pm \sqrt{10}$
2 $\frac{3}{2}$
3 $-\frac{7}{6} \pm \frac{\sqrt{73}}{6}$
4 $\frac{4}{11} \pm \frac{\sqrt{284}}{22}$ or $\frac{4}{11} \pm \frac{\sqrt{71}}{11}$

Roots of equations and use of the discriminant

1 $\Delta = 121 \Rightarrow$ 2 real roots
2 $\Delta = 0 \Rightarrow$ 1 real root
3 $\Delta = -31 \Rightarrow$ no real roots
4 $\Delta = -7 \Rightarrow$ no real roots
5 $p > 3.0625$
6 $p > 19.90$ or $p < -19.90$
7 $k < 8$
8 $d \leq -1$ or $d \geq 11$

2 Circles (pp. 12–14)

1 $(x + 7)^2 + (y - 3)^2 = 9$ G
$(x + 5)^2 + (y + 3)^2 = 9$ C
$(x - 3)^2 + (y + 7)^2 = 9$ E
$(x - 7)^2 + (y + 3)^2 = 9$ A
$(x - 5)^2 + (y - 7)^2 = 9$ B
$(x + 3)^2 + (y - 5)^2 = 9$ D
$(x - 7)^2 + (y - 5)^2 = 9$ F

2 $(x + 6)^2 + (y - 2)^2 = 4$ A
$x^2 + (y - 2)^2 = 1$ C
$(x - 2)^2 + (y + 6)^2 = 9$ G
$(x + 2)^2 + y^2 = 1$ D
$(x - 2)^2 + (y - 6)^2 = 9$ B
$(x + 2)^2 + (y + 6)^2 = 4$ H
$x^2 + (y + 2)^2 = 1$ J
$(x - 6)^2 + (y + 2)^2 = 9$ F
$(x - 2)^2 + y^2 = 1$ I
$(x - 6)^2 + (y - 2)^2 = 4$ E

3 Trigonometry (pp. 15–22)

Radians

1 π
2 $\frac{\pi}{2}$
3 $\frac{\pi}{6}$
4 $\frac{5\pi}{6}$
5 60°
6 45°
7 135°
8 120°
9

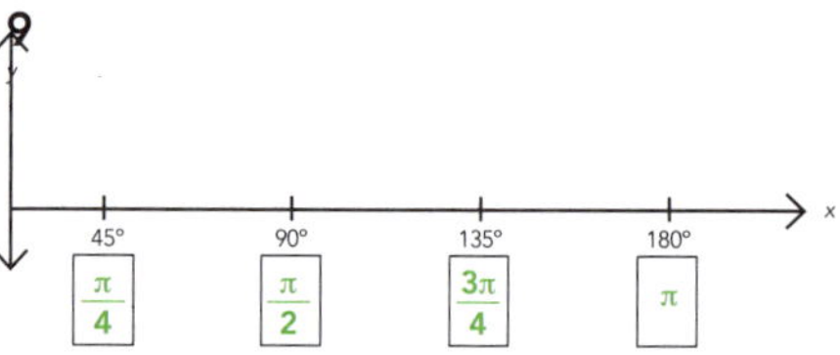

10

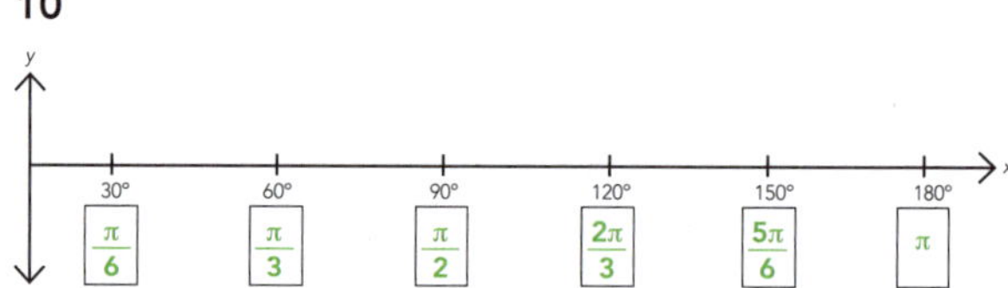

Putting it together

1 0
2 1
3 1
4 –1
5 0.5
6 –1
7 0.7071 or $\frac{1}{\sqrt{2}}$
8 –0.8660 or $\frac{-\sqrt{3}}{2}$
9 $\theta = 60° = \frac{\pi}{3}$ or $\theta = -60° = -\frac{\pi}{3}$
10 $\theta = 120° = \frac{2\pi}{3}$ or $\theta = -120° = -\frac{2\pi}{3}$
11 $\theta = 45° = \frac{\pi}{4}$ or $\theta = -135° = -\frac{3\pi}{4}$
12 $\theta = 60° = \frac{\pi}{3}$ or $\theta = -60° = -\frac{\pi}{3}$
13 $\theta = 150° = \frac{5\pi}{6}$ or $\theta = -150° = -\frac{5\pi}{6}$
14 $\theta = 90° = \frac{\pi}{2}$

Surds (pp. 24–34)

Rules for surds (pp. 24–26)

1 9
2 3
3 2
4 4
5 xy^3
6 $2x^2yz^4$
7 $5p^2q^5\sqrt{3pr}$
8 $\frac{4q^3\sqrt{5}}{5p}$
9 $2\sqrt{3}$
10 $3\sqrt{5}$
11 $6\sqrt{7}$
12 $12\sqrt{3}$
13 $7p\sqrt{q}$
14 $\frac{y\sqrt{x}}{2}$ or $\frac{y}{2}\sqrt{x}$
15 $9\sqrt{3}$
16 $5\sqrt{3}$
17 $72\sqrt{2}$
18 $\frac{\sqrt{7}}{3}$
19 $10a^2b^2\sqrt{6}$
20 $2b^2\sqrt{3ab}$

ISBN: 9780170447058

21 $\frac{\sqrt{11}}{3\sqrt{3}}$ 22 $\sqrt{\frac{7p}{q}}$
23 $\frac{7\sqrt{2}}{11\sqrt{5}}$ 24 $\frac{9\sqrt{5}}{5\sqrt{21}}$

Simplifying expressions containing surds (p. 27)
1 $\sqrt{14}+\sqrt{21}$ 2 $5\sqrt{6}-6$
3 $\sqrt{6}+6\sqrt{3}-5\sqrt{2}-30$ 4 4
5 $8-2\sqrt{7}$ 6 $3x-x\sqrt{y}$ or $x(3-\sqrt{y})$
7 $p-2\sqrt{pq}+q$ 8 $4x+4\sqrt{3xy}+3y$
9 $15-14x-11\sqrt{x}$ 10 $\sqrt{x}(x+12)-6x-8$

Rationalising and more simplifying of expressions containing surds (pp. 28–30)
1 $\frac{\sqrt{6}}{2}$ 2 $8\sqrt{x}$
3 $\frac{5\sqrt{p}}{2p}$ 4 $\frac{3\sqrt{2xy}}{y}$
5 $\frac{3\sqrt{a-b}}{a-b}$ 6 $\frac{\sqrt{2x-1}}{6x-3}$
7 $\sqrt{3}-1$ 8 $2(3+\sqrt{7})$
9 $10-2\sqrt{7}$ 10 $14+2\sqrt{5}$
11 $-3-\sqrt{3}$ 12 $\sqrt{7}+3$
13 $\frac{1+\sqrt{a}}{1-a}$ 14 $\frac{6p-2\sqrt{p}}{9p-1}$
15 $8+6\sqrt{3}$ 16 $\frac{1+5\sqrt{5}}{4}$
17 $\frac{1+3\sqrt{x}}{x-1}$ 18 $\frac{22+5\sqrt{y}-3y}{4-y}$

Equations involving surds (pp. 31–34)
1 $x=6$ (but not 1) 2 $x=1$ (but not –6)
3 $x=3$ (but not 1) 4 $x=10$ (but not –2)
5 $x=0$ or 5 6 $x=7$ or –1
7 $x=\frac{(k+16)^2}{64}$ 8 $x=\frac{(121+5k)^2}{484}$
9 $x=\frac{(9-2k)^2}{36}$ 10 $x=\frac{(1-7k)^2}{4}$
11 $x=\frac{k^2}{1-k^2}$ 12 $x=\frac{4k^2}{9-k^2}$
13 $x=\frac{63}{16k^2-9}$ 14 $x=\frac{5+k^2}{2k^2-1}$

Polynomials (pp. 35–51)

Completing the square (pp. 35–38)
1 $x^2+4x+4=(x+2)^2$ 2 $x^2-10x+25=(x-5)^2$
3 $x^2-8x+16=(x-4)^2$ 4 $x^2+24x+144=(x+12)^2$
5 $2(x^2+6x+9)=2(x+3)^2$ 6 $3(x^2-14x+49)=3(x-7)^2$
7 –7, 5 8 3, –1
9 –2, 9 10 $6\pm\sqrt{13}$
11 $4\pm\sqrt{10}$ 12 $-1.5\pm\sqrt{1.5}$
13 2.5, –6.5 14 5, –3.5
15 $-\frac{1}{2}\pm\sqrt{\frac{5}{3}}$ 16 $\frac{3}{4}\pm\sqrt{\frac{17}{16}}$
17 $(x-1)^2+1$ 18 $(x-3)^2+1$
19 $(x-5)^2+5$ 20 $\left(x-\frac{1}{2}\right)^2+4\frac{3}{4}$
21 $2(x-2)^2+3$ 22 $3(x-4)^2+1$
23 $5(x-3)^2+2$ 24 $9(x-4)^2+1$

Long division (pp. 39–44)
1 $x^2+7x+10$ 2 x^2+4x+9
3 $4x^2+10x+5$ 4 $x^2-9x+11$
5 $2x^2+x-6$ 6 x^3-2x^2-x+2
7 $x^2-2x-3+\frac{4}{x-2}$ 8 $2x^2+3x-10-\frac{5}{x-3}$
9 $(2x-1)(x-4)(3x-1)$ 10 $(x-6)(x+4)(x+2)$
11 $(3x-2)(2x-1)(2x+3)$ 12 $(3x-2)^3$
13 $a=2, b=3, c=-10, d=6$
14 $a=-3, b=3, c=-15, d=20$
15 $2x-1$ 16 $2x^2+5x+3$

Factor and remainder theorems (pp. 45–51)

1 The factor theorem
1 $(x-1)(x+1)(x-2)$ 2 $(x+1)(x-2)(x+2)$
3 $(x+2)^2(x-3)$ 4 $(2x-1)(x+3)(x-2)$
5 $(3x-1)(x+2)(x+4)$ 6 $(4x-1)(x+3)(x-2)$
7 $(2x+3)(x+5)(x-1)$ 8 $(6x-1)(x+3)(x+8)$

2 The remainder theorem
9 8 10 –36
11 20 12 –48
13 1 14 –25
15 11 16 –20
17 $p=-42$, remaining solutions are 2 and $\frac{1}{5}$
18 $p=-3$ and $q=-17$
19 $p=-11$ and $q=-26$

Complex numbers (pp. 52–115)

Imaginary numbers (pp. 52–53)
1 $10i$ 2 $i(\sqrt{8}-5)$
3 –26 4 3
5 $-i$ 6 i
7 $-128i$ 8 $-\frac{1}{i}$

Complex numbers — the basics (pp. 54–55)
1 $\text{Re}(z)=2$, $\text{Im}(z)=9$ 2 $\text{Re}(z)=27$, $\text{Im}(z)=0$
3 $\text{Re}(z)=0$, $\text{Im}(z)=15$ 4 $\text{Re}(z)=x$, $\text{Im}(z)=-y$
5 $\text{Re}(z)=4$, $\text{Im}(z)=-(x+2)$ 6 $\text{Re}(z)=6-2y$, $\text{Im}(z)=1+x$
7 $\text{Re}(z)=9$, $\text{Im}(z)=13-9x$ 8 $\text{Re}(z)=3x$, $\text{Im}(z)=1-2y$
9 $x=-5$, $y=-2$ 10 $x=-1$, $y=5$
11 $x=0$, $y=13$ 12 $x=-2$, $y=-6$
13 $x=0$, $y=1$ 14 $x=3$, $y=-2$
15 $x=-5$, $y=1$ 16 $x=2$, $y=-3$

Complex numbers and Argand diagrams (p. 56)

A: $4 + 9i$ **B**: $1 - 3i$
C: $-8 + 5i$ **D**: $-3 - 8i$
E: 7

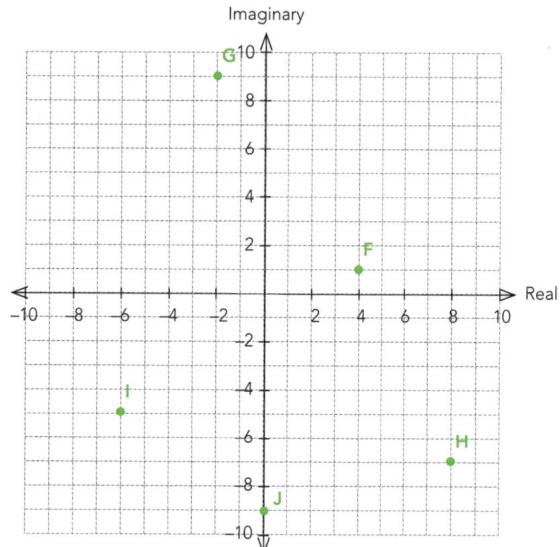

Manipulation of complex numbers (pp. 57–64)

Addition and subtraction

1 $7 + 9i$ **2** $3 - 5i$
3 $-3 - i$ **4** $-2 - 9i$
5 $7 - 2i$ **6** $-2 - 2i$
7 $-1 + 11i$ **8** $1 - 5i$
9 $(x + y) + (y - x)i$ **10** $2x + (1 - 2y)i$
11 $3x - yi$ **12** $(x + 2y) + (3y + x)i$
13

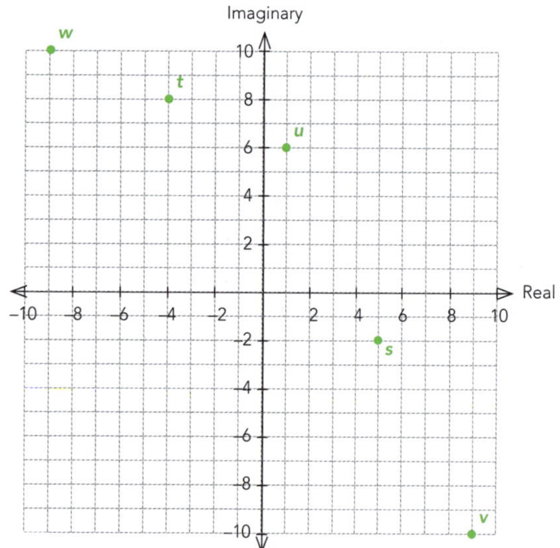

Multiplication

1 $10 - 15i$ **2** $-3 + 6i$
3 $4 + 5i$ **4** $-6 - 21i$
5 $-5 + 14i$ **6** $20 - 9i$
7 $26 - 13i$ **8** $4 - 17i$
9 $-29 - 11i$ **10** $6 + 27i$
11 $-5 + 12i$ **12** $-16 - 30i$
13 $(x^2 + y^2)i$ **14** $14x + (3x^2 - 8)i$
15 $-xy - 13xyi$ **16** $(2y - 3x^2) + x(3 + 2y)i$
17 6 **18** $2 + 11i$
19 -8 **20** 51
21 $b = 2$ **22** $b = \sqrt{3}$
23 $x = 5, y = 3$

Division

1 $\bar{p} = 2 - 7i$ **2** $\bar{s} = 6 + 11i$
3 $q \times \bar{q} = 5$ **4** $r \times \bar{r} = 34$
5 $a = \pm 2$ **6** $b = \pm 3$
7 $\frac{2}{13} - \frac{3}{13}i$ **8** $\frac{20}{29} + \frac{8}{29}i$
9 $\frac{2}{13} + \frac{3}{13}i$ **10** $-\frac{5}{13} + \frac{1}{13}i$
11 $-\frac{1}{29} + \frac{17}{29}i$ **12** $1 - 2i$
13 $-\frac{17}{13} - \frac{19}{13}i$ **14** $\frac{11}{50} + \frac{23}{50}i$
15 $k = 1$ **16** $k = -1$
17 $p = ac + bd$
$q = bc - ad$
18 **a** $-i$

b
$$\frac{k - i}{1 + ki} = \frac{k - i}{1 + ki} \times \frac{1 - ki}{1 - ki}$$
$$= \frac{k - k^2 i - i + ki^2}{1 + k^2}$$
$$= \frac{k - i(k^2 + 1) - k}{1 + k^2}$$
$$= -i$$

19
$$\frac{k + 9i}{1 + ki} = \frac{k + 9i}{1 + ki} \times \frac{1 - ki}{1 - ki}$$
$$= \frac{k - k^2 i + 9i - 9ki^2}{1 + k^2}$$
$$= \frac{10k}{1 + k^2} + \frac{9 - k^2}{1 + k^2}i$$
$$\therefore \frac{9 - k^2}{1 + k^2} = 0$$
$$9 - k^2 = 0 \Rightarrow k = \pm 3$$

Mixing it up (pp. 65–67)

1 **a** $5 - 10i$
b $p = 5$ and $q = 3$
2 **a** $-5 + 13i$
b $p = 7$ and $q = 3$
3 $\frac{2}{5} + \frac{7}{10}i$
4 $2\frac{2}{5} - 4\frac{4}{5}i$
5 $-7\frac{9}{10} + 6\frac{3}{10}$
6 $-\frac{1}{5} + \frac{4\sqrt{3}}{5}i$ or equivalent
7 $-4\frac{164}{169} + 11\frac{157}{169}i$
8 $5 + 15i$
9 $3 - 4i$
10 $a = -7$ and $b = -24$

ISBN: 9780170447058

11 Let $u = a + bi$ and $v = c + di$.

$uv = (a + bi)(c + di)$
$= ac + adi + bci + bdi^2$
$= (ac + bd) + (ad + bc)i$
$\mathbf{\overline{uv} = (ac + bd) - (ad + bc)i}$
$\bar{u} \times \bar{v} = (a - bi) \times (c - di)$
$= ac - adi - bci + bdi^2$
$\mathbf{\bar{u} \times \bar{v} = (ac - bd) - (ad + bc)i}$
$\therefore \mathbf{\overline{uv} = \bar{u} \times \bar{v}}$

The modulus (pp. 68–70)

1 $\sqrt{13}$ **2** 6
3 5 **4** 2
5 $\sqrt{73}$ **6** $\sqrt{x^2 + 4x + 13}$
7 $\sqrt{13}$ **8** $\sqrt{26}$
9 $2\sqrt{5}$ **10** $\sqrt{73}$
11 $\frac{1}{\sqrt{13}}$ **12** $\frac{1}{\sqrt{2}}$
13 2.55 **14** 0.5
15 $5\sqrt{2}$ **16** 10

17 **a** $|z| = |(a^2 + b^2)|$
$= \sqrt{a^2 + b^2}$
$|\bar{z}| = |(a^2 - b^2)|$
$= \sqrt{a^2 + b^2}$
$\therefore |z| = |\bar{z}|$

b z and $\bar{z}$ are mirror images of each other in the x-axis. $|z|$ and $|\bar{z}|$ both measure the length of the vector from the origin to z and $\bar{z}$, so they must be the same length.

18 $|z|^2 = |a + bi|^2$
$= (\sqrt{a^2 + b^2})^2$
$= a^2 + b^2$
$z\bar{z} = (a + bi)(a - bi)$
$= a^2 + b^2$
$\therefore |z|^2 = z\bar{z}$

19 $|zw| = |(a + bi)(c + di)|$
$= |(ac + adi + bci - bd)|$
$= |(ac - bd) + (ad + bc)i|$
$= \sqrt{(ac - bd)^2 + (ad + bc)^2}$
$= \sqrt{a^2c^2 - 2abcd + b^2d^2 + a^2d^2 + 2abcd + b^2c^2}$
$= \sqrt{a^2c^2 + b^2d^2 + a^2d^2 + b^2c^2}$
$|z||w| = \sqrt{a^2 + b^2} \times \sqrt{c^2 + d^2}$
$= \sqrt{(a^2 + b^2) \times (c^2 + d^2)}$
$= \sqrt{a^2c^2 + b^2d^2 + a^2d^2 + b^2c^2}$
$\therefore |zw| = |z||w|$

Polynomials again (pp. 71–80)

Quadratics

1 $\pm 10i$ **2** $\pm 5i$
3 $1 \pm 2i$ **4** $2 \pm i$
5 $-2 \pm 2i$ **6** $1 \pm \sqrt{3}i$
7 $3 \pm \sqrt{3}i$ **8** $-\frac{1}{2} \pm \frac{\sqrt{3}}{2}i$
9 $(x - 3 - 5i)(x - 3 + 5i)$
10 $\left(x + \frac{1}{3} + \frac{\sqrt{2}}{3}i\right)\left(x + \frac{1}{3} - \frac{\sqrt{2}}{3}i\right)$
11 Other solution: $z = 2 - 3i$
a = 1, b = –4, c = 13
12 Other solution: $z = 3 - \sqrt{3}i$
a = 1, b = –6, c = 12
13 Other solution: $z = -1 - \sqrt{2}i$
a = 1, b = 2, c = 3

Cubics

1 A = –2. Other two factors: 4, $-1 + 2i$
2 A = 1. Other two factors: –2, $3 - 2i$
3 A = –13. Other two factors: 1, $6 + 2i$
4 A = 9. Other two factors: 2, $-4 + 3i$
5 A = 11. Other two factors: $-\frac{1}{3}$, $2 + i$
6 A = –19. Other two factors: $-\frac{1}{5}$, $2 - i$
7 A = 16. Other two factors: $\frac{3}{2}$, $1 + 2i$
8 $2, -2 \pm 5i$ **9** $-1, 5 \pm 2i$
10 $-1, 1 \pm i$ **11** $2, -1 \pm \sqrt{3}i$

Polar coordinates (pp. 81–88)

Conversion from Cartesian form to polar form

1 $z = \sqrt{2}\text{ cis } 45°$ or $\sqrt{2}\text{ cis }\frac{\pi}{4}$
2 $z = \text{cis }(-135°)$ or $\sqrt{2}\text{ cis}\left(-\frac{3\pi}{4}\right)$
3 $z = 5\text{ cis } 143.1°$ or $5\text{ cis } 2.498$
4 $z = 2\text{ cis }(-60°)$ or $2\text{ cis}\left(-\frac{\pi}{3}\right)$
5 $z = \sqrt{74}\text{ cis }(-35.5°)$ or $\sqrt{74}\text{ cis }(-0.6202)$
6 $z = \sqrt{41}\text{ cis }(-128.7°)$ or $\sqrt{41}\text{ cis }(-2.245)$
7 $z = \sqrt{5}\text{ cis } -26.6°$ or $\sqrt{5}\text{ cis } -0.4636$
8 $z = \sqrt{109}\text{ cis } -73.3°$ or $\sqrt{109}\text{ cis }(-1.2793)$

Conversion from polar form to Cartesian form

9 $z = 3 + 4i$ **10** $z = -1 + i$
11 $z = -1 - \sqrt{3}i$ **12** $z = 1 - 10i$
13 $z = -3 + 3i$ **14** $z = 3 + 4i$
15 $z = -2 + 11i$

Multiplying and dividing complex numbers in polar form

1 **a** 36 cis 1.4 **b** 4 cis (–0.2)
2 **a** $20\text{ cis }\frac{3\pi}{4}$ **b** $5\text{ cis }\frac{7\pi}{12}$
3 **a** 48 cis 0.637 **b** 0.75 cis (–0.187)
4 **a** $3p^2\text{ cis } 3q$ **b** $3\text{ cis }(-q)$

Mixing it up (pp. 89–92)

1 $u\bar{u} = (a + bi)(a - bi)$
$= a^2 - b^2 i^2$
$= a^2 + b^2$
$\therefore u\bar{u} = |u|^2$

$|\bar{u}|^2 = (\sqrt{a^2 + b^2})^2$
$= a^2 + b^2$

2 $a = -0.5$
$b = 0.5$
$\arg(w) = \frac{3\pi}{4}$

3 $3\text{ cis}\left(-\frac{\pi}{2}\right)$ **4** $w = 10\text{ cis}\left(-\frac{\pi}{2}\right)$

5 **a** $p^4\text{ cis }\frac{13\pi}{20}$ **b** $p^2\text{ cis }\frac{3\pi}{20}$
c $p^6\text{ cis }\frac{4\pi}{5}$

6 $u = i$

7 $w = \sqrt{3} + \sqrt{3}\,i$

8 $w = -2 + 2i$

9 $w = 2 + i$ or $w = 2 + 3i$

10 $w = 1 + 2i$ or $w = -\dfrac{1}{3} + 2i$

11 $z = -1.6 + 0.8i$
$\arg(z) = 2.678$

12 a $\dfrac{1+2i}{p+qi} \times \dfrac{p-qi}{p-qi} = \dfrac{(p+2q)+i(2p-q)}{p^2+q^2}$

$\arg(z) = \dfrac{\pi}{4} \Rightarrow p + 2q = 2p - q$

$\therefore\ 3q = p$

b $q = 1, p = 3$

$z = \dfrac{1+2i}{3+i} = 0.5 + 0.5i$

13 0.4472 cis (–2.677)

14 8 cis (–π)

Powers of complex numbers in polar form: de Moivre's theorem (pp. 93–95)

1 64 cis 0.6 2 $16 \text{ cis } \dfrac{4\pi}{5}$

3 $27 \text{ cis}\left(-\dfrac{3\pi}{4}\right)$ 4 $32 \text{ cis } \dfrac{3\pi}{4}$

5 $4 \text{ cis}\left(\dfrac{2\pi}{3}\right)$ 6 0.125 cis 1.217

7 $-8 + 8\sqrt{3}i$

8 $5832i = 5832 \text{ cis } \dfrac{\pi}{2}$

9 $120 - 34.93i$

10 $z = \dfrac{\sqrt{3}}{2} \text{ cis } \dfrac{\pi}{6}$ $z^2 = \dfrac{3}{4} \text{ cis } \dfrac{\pi}{3}$

$z^3 = \dfrac{3\sqrt{3}}{8} \text{ cis } \dfrac{\pi}{2}$ $z^4 = \dfrac{9}{16} \text{ cis } \dfrac{2\pi}{3}$

$z^5 = \dfrac{9\sqrt{3}}{32} \text{ cis } \dfrac{5\pi}{6}$ $z^6 = \dfrac{27}{64} \text{ cis } \pi$

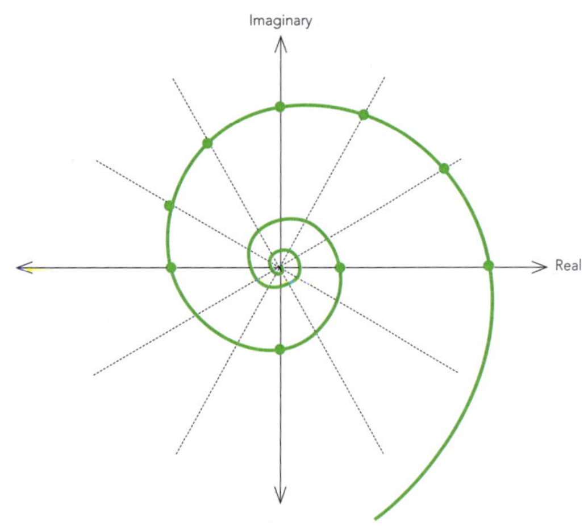

Solving polynomial equations using de Moivre's theorem (pp. 96–102)

1 $\sqrt{2} \text{ cis } \dfrac{\pi}{6}, \sqrt{2} \text{ cis}\left(-\dfrac{5\pi}{6}\right)$

2 $2, 2 \text{ cis } \dfrac{2\pi}{3}, 2 \text{ cis}\left(-\dfrac{2\pi}{3}\right)$

3 $-2, -2 \text{ cis } \dfrac{\pi}{5}, 2 \text{ cis } \dfrac{3\pi}{5}, -2 \text{ cis}\left(-\dfrac{\pi}{5}\right), -2 \text{ cis}\left(-\dfrac{3\pi}{3}\right)$

4 $\text{cis}\left(-\dfrac{\pi}{4}\right)$ and $\text{cis } \dfrac{3\pi}{4}$

5 $\sqrt[6]{2} \text{ cis } \dfrac{\pi}{4}, \sqrt[6]{2} \text{ cis } \dfrac{11\pi}{12}, \sqrt[6]{2} \text{ cis}\left(-\dfrac{5\pi}{12}\right)$

6 $a, a \text{ cis } \dfrac{2\pi}{3}, a \text{ cis}\left(-\dfrac{2\pi}{3}\right)$

7 $2 \text{ cis } \dfrac{2\pi}{9}, 2 \text{ cis } \dfrac{8\pi}{9}, 2 \text{ cis}\left(-\dfrac{4\pi}{9}\right)$

8 $\sqrt[5]{k} \text{ cis } \dfrac{\pi}{10}, \sqrt[5]{k} \text{ cis } \dfrac{\pi}{2}, \sqrt[5]{k} \text{ cis } \dfrac{9\pi}{10}, \sqrt[5]{k} \text{ cis}\left(-\dfrac{7\pi}{10}\right),$
$\sqrt[5]{k} \text{ cis}\left(-\dfrac{3\pi}{10}\right)$

9 $\sqrt{3p} \text{ cis } \dfrac{\pi}{4}, \sqrt{3p} \text{ cis } \dfrac{3\pi}{4}, \sqrt{3p} \text{ cis}\left(-\dfrac{3\pi}{4}\right), \sqrt{3p} \text{ cis}\left(-\dfrac{\pi}{4}\right)$

Finding the locus of complex numbers subject to restrictions (pp. 103–115)

1 Where the real and/or imaginary part of the complex number is restricted

1 $y = 2$

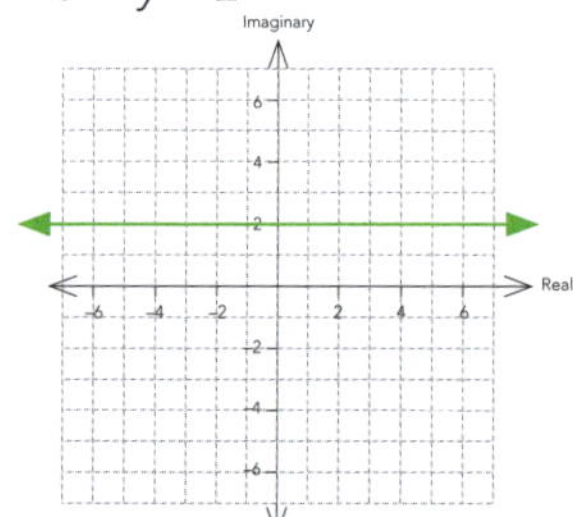

2 $x = -4$

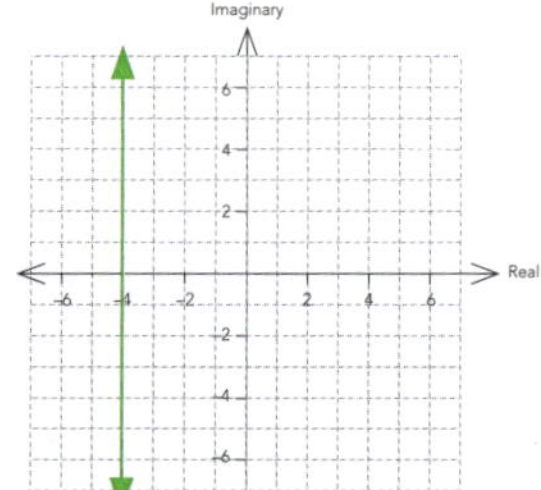

3 $x = 4$

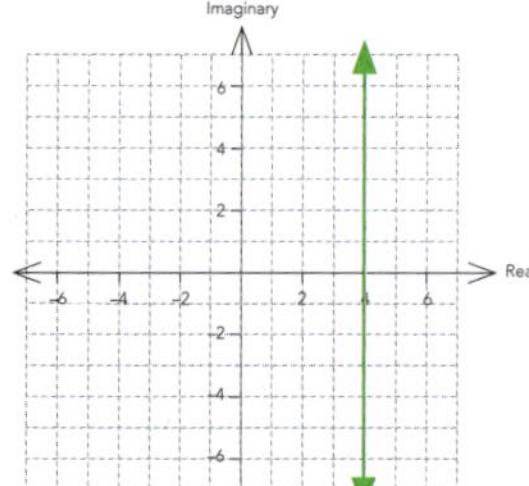

4 $y = x - 1$

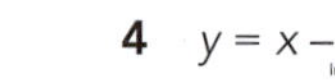

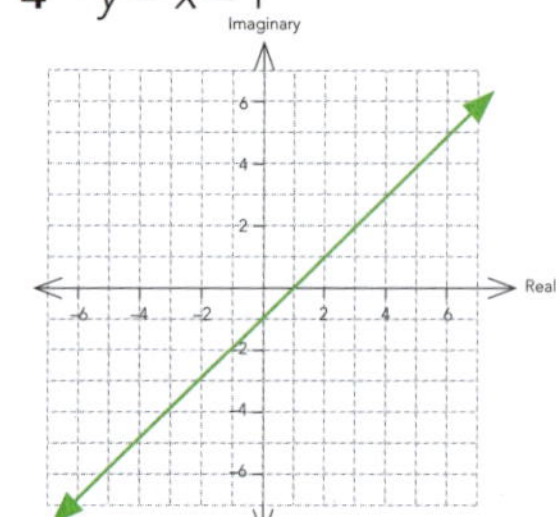

5 $y = -x + 5$

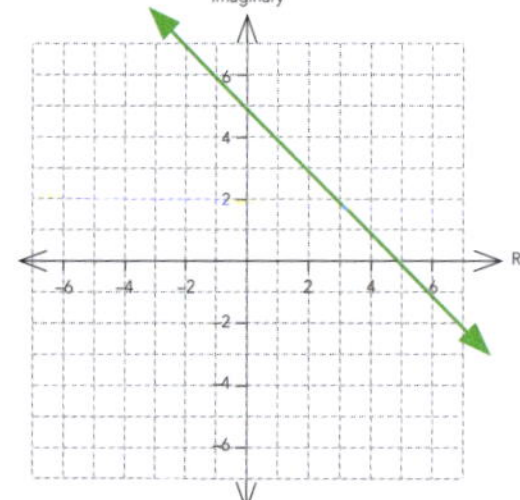

6 $y = -2x + 7$

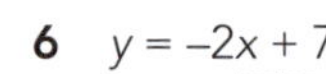

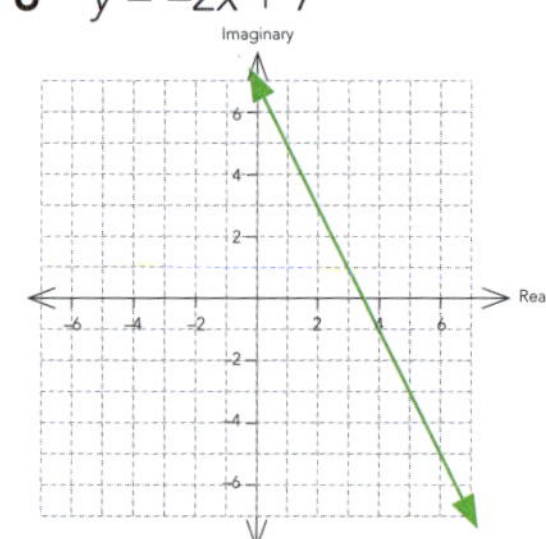

7 $y = -\dfrac{3}{2}x - 6$

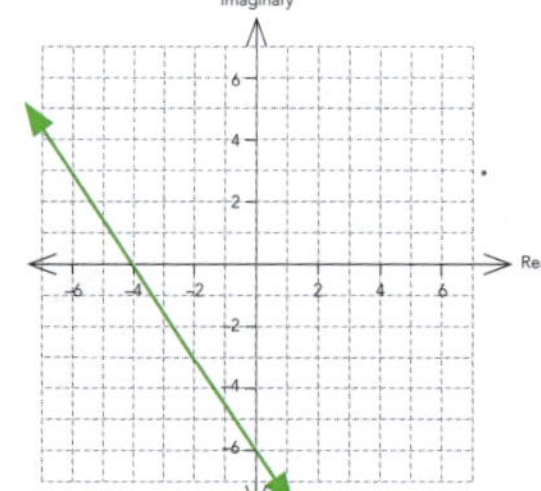

8 $x - y \leq 6$

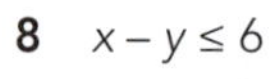

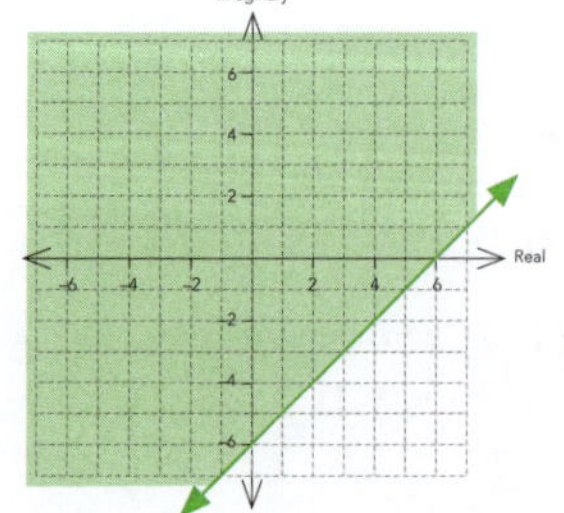

 ISBN: 9780170447058

2 Where the modulus of the complex number is restricted

1 **a** $x^2 + y^2 = 25$ **b** $x^2 + (y + 3)^2 = 16$

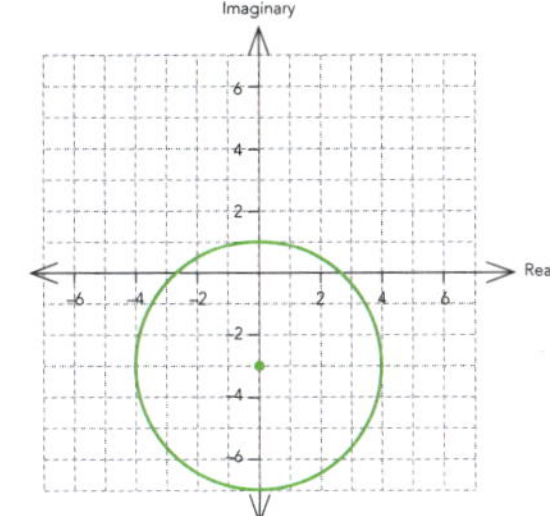

2 $(x - 5)^2 + y^2 = 9$
3 $(x - 1)^2 + (y + 2)^2 = 49$
4 gradient = –3
5 $y = 5x - 2$
6 $m = 6$
7 $(x - 1)^2 + (y + 1)^2 = 8$
8 $(x + 2)^2 + (y - 6)^2 = 40$

3 Where the argument of the complex number is restricted

1 $y = -x$, where $x > 0$, $y < 0$.

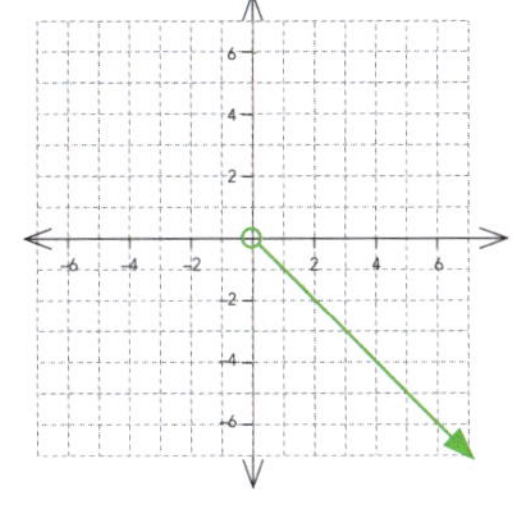

2 $y = -\frac{1}{\sqrt{3}}x$, where $x < 0$ and $y > 0$.

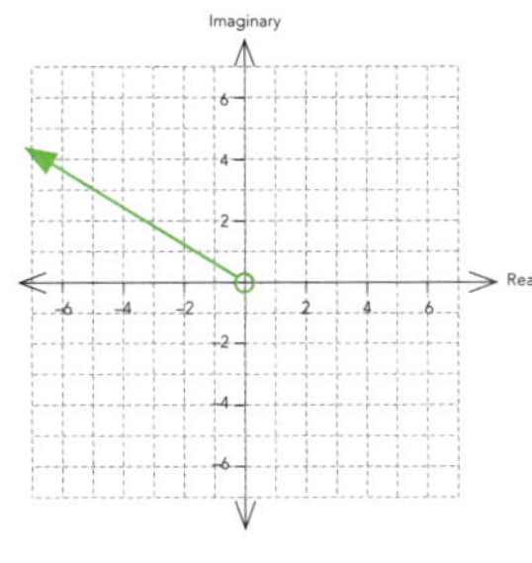

3 $y = 2 - x$, where $x > 2$ and $y < 0$.

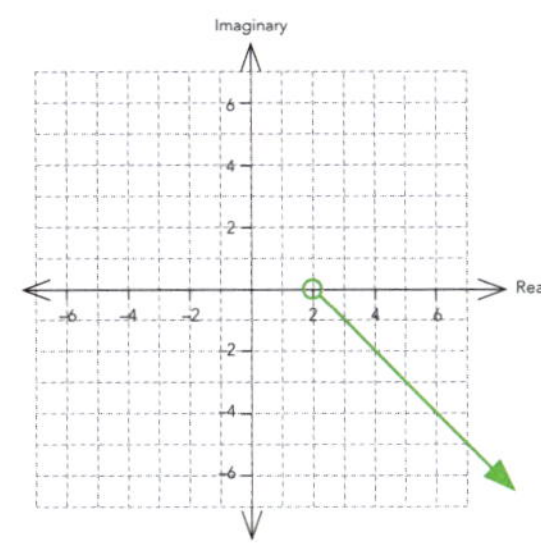

4 $y = \sqrt{3}x - 1$, where $x > 0$ and $y > -1$.

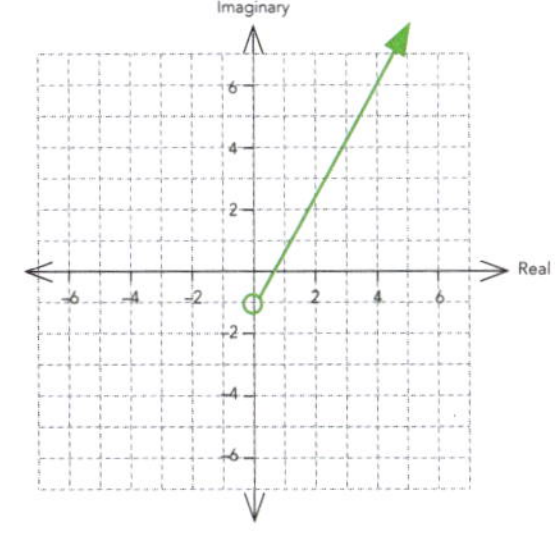

5 $y = x + 5$, where $x > -2$ and $y > 3$.

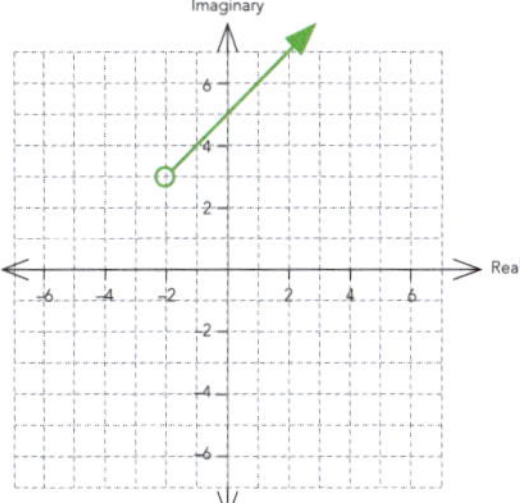

6 $y = \frac{1}{\sqrt{3}}x - 4.5774$, where $x > 1$ and $y > -4$.

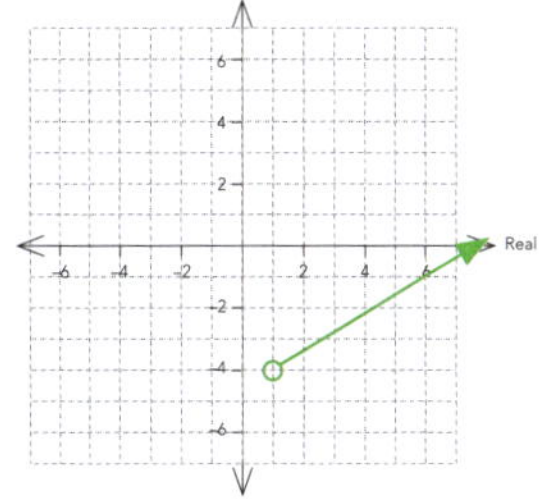

7 $x^2 + y^2 - x + y = 0$
8 $x^2 + y^2 - 2x - 4y - 3 = 0$
9 $x^2 + y^2 + 3x + 5y - 4 = 0$
10 $x^2 + y^2 + 7x - 7y + 12 = 0$
11 $x^2 + y^2 - 5x + 5y + 6 = 0$

Practice questions (pp. 116–121)

Practice question one (pp. 116–117)

a $\frac{u}{v} = p^3 \text{ cis } \frac{5\pi}{12}$

b m = –1

c $x = \frac{(25 - 2k)^2}{100}$

d $|\overline{w}| = 5$
$v = 0.1 - 0.3i$

e $z^3 = 2p \text{ cis}\left(\frac{\pi}{4}\right)$

$z_n = (2p)^{\frac{1}{3}} \text{ cis}\left(\frac{2n\pi}{3} + \frac{\pi}{12}\right)$

$z_1 = \sqrt[3]{2p} \text{ cis}\left(\frac{\pi}{12}\right)$

$z_2 = \sqrt[3]{2p} \text{ cis}\left(\frac{9\pi}{12}\right)$

$z_3 = \sqrt[3]{2p} \text{ cis}\left(\frac{17\pi}{12}\right) = \sqrt[3]{2p} \text{ cis}\left(-\frac{7\pi}{12}\right)$

Practice question two (pp. 118–119)

a

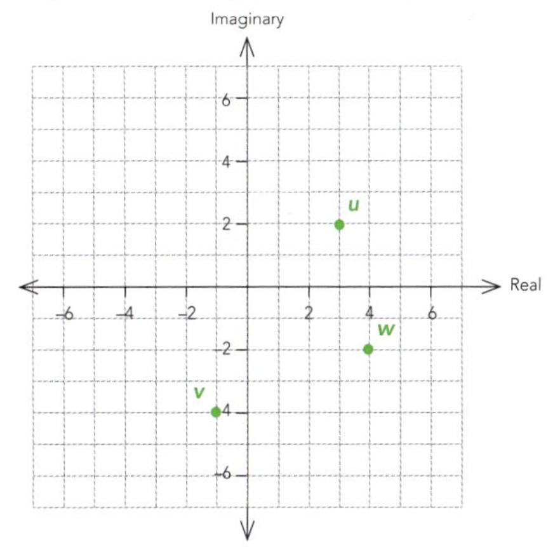

ISBN: 9780170447058

b $\arg(u) = \frac{-3\pi}{4}$ or $-135°$

c $k = \pm 6$

d A = 7, B = 3 and C = –5

e $y = -x - 3$, where $x < 2$ and $y > -5$.

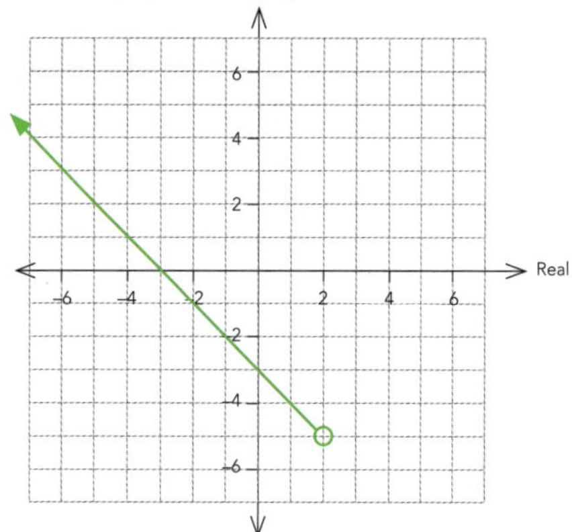

Practice question three (pp. 120–121)

a $2 - 0.5\sqrt{6}$

b 2500 cis π or 2500 cis 180°

c A = –17

Two other solutions: $z = 3 - 2i$ and $z = -5$.

d $p(9p - 10) < 0 \Rightarrow 0 < p < \frac{10}{9}$

e $(x - 2)^2 + (y + 1)^2 = 8$

This represents a circle with a centre at (2, –1) and a radius of $\sqrt{8}$.

 ISBN: 9780170447058